T0168796

GROUND FORCES
for a
RAPIDLY EMPLOYABLE
JOINT TASK FORCE

First-Week Capabilities for Short-Warning Conflicts

EUGENE C. GRITTON · PAUL K. DAVIS
RANDALL STEEB · JOHN MATSUMURA

Prepared for the
Office of the Secretary of Defense
United States Army

**National Defense Research Institute
Arroyo Center**

RAND

Approved for public release; distribution unlimited

This monograph is a think piece about rapidly deployable future ground forces that would be used in time-urgent joint-task-force missions. We sketch a vision of what is needed operationally, describe forces responsive to those needs, and then discuss the feasibility of achieving such forces—first with near-term systems but new operational concepts, and then, for the longer term, by drawing upon advanced technology often discussed under the rubric of the revolution in military affairs or RMA. Thus, our proposals call for a *vigorous evolution* with tangible accomplishments within five years, rather than a discontinuous revolution in 10 or 20 years. Our analysis suggests that such forces would potentially be quite valuable, while also indicating their likely limitations and the many uncertainties of our assessment.

The monograph grew out of our efforts during the 1998 Defense Science Board summer study, a study for the Department of Defense on "transforming the force," and a small study for the Joint Staff on strategic mobility. Those efforts, in turn, drew on RAND work for the Army, the Defense Advanced Research Projects Agency (DARPA), Air Force, and Joint Staff. In preparing this integrative study, we drew heavily on work initially performed in RAND's Army-sponsored Arroyo Center. As a result, this monograph is being jointly published by NDRI and the Arroyo Center. The monograph is intended for a broad audience interested in future ground forces in a joint-task-force context.

Our work has been accomplished in the Acquisition and Technology Policy Center of RAND's National Defense Research Institute, a federally funded research and development center (FFRDC) sponsored by the Office of the Secretary of Defense, the Joint Staff, the unified commands, and the defense agencies.

CONTENTS

FIGURES

TABLES

OBJECTIVES

In many military interventions to deter or thwart an invader entering a country friendly to the United States, the United States would benefit greatly from being able to employ—within *days* rather than weeks—a joint task force that would combine long-range fires from aircraft and missiles with maneuver forces on the ground. Although long-range fires alone might be sufficient in some cases, in many others they would not. If suitable maneuver forces existed and could be used (a function of circumstances), they could greatly enhance the effectiveness of long-range fires, further damage and disrupt the enemy themselves, hedge against some potential failures of the fires, and accomplish other important objectives. They could be especially valuable for conflicts in mixed terrain, and in conflicts in which they linked up with significant friendly forces. They could be quite useful in smaller-scale contingencies as well as major wars. For example, had such forces existed in 1999, more options would have been available to NATO's political leaders in the early stages of the Kosovo operation.

This monograph is a think piece describing a concept for such a ground-force component of a rapidly deployable joint task force. Our objective is to provide a coherent and relatively concrete vision of what could be accomplished with available technology and technology that is reasonably achievable. We believe that much of the vision is sound, and that the rest of it will at least provide a serious straw man from which to depart. We also hope that our work will add balance to the discussion of advanced forces. While we and many others have studied the potential of long-range fires for some years, the literature contains much less analysis on how ground forces could contribute early in a conflict—notably during the "halt phase," which is often conceived solely in terms of interdiction with aircraft and missiles.

THE TIMELINE CHALLENGE

Given likely U.S. capabilities, future aggressors will have incentives to act with little actionable warning and to maneuver their forces as fast as possible to occupy cities, capture key installations, and reduce their vulnerability to U.S. responses. The United States should therefore be prepared to disrupt an invader within the first days of invasion, and to thwart the invasion altogether within perhaps a week. Otherwise, the invader could often achieve major objectives before the United States reacted, after which military and political factors might make response difficult and impractical. That is, forward defense is often very important strategically—even if difficult and inconvenient militarily. Unfortunately, the United States should assume that warning will be ambiguous until D-Day itself and that the invader will use or threaten use of weapons of mass destruction, thereby endangering equipment in fixed locations and deployments into some of the major bases and ports.

It is for this context that we considered concepts for achieving first-week ground-force capabilities in joint-task-force operations. Except in instances (such as today's Korea) where appropriate forces are already in place, meeting the challenge will require changes in strategic mobility, forces, and doctrine. Since forward stationing of ground forces has high political and other costs—for both host countries and the United States—it seems prudent to assume the need to be able to rapidly deploy ground forces from the sea or from the United States.

WHAT IS FEASIBLE WITH STRATEGIC MOBILITY?

Recent discussion of rapid deployment concepts has often postulated dramatic improvements in airlift and sealift, while also postulating greatly lightened forces deployable by airlift. Our own conclusions are rather different:

- Truly light ground forces will be insufficient because technology will not soon provide protection against even such ubiquitous weapons as rocket launchers and cannons. Instead, a *combination* of light mobile infantry and light (or medium-weight) mechanized forces will be needed early.

- For the foreseeable future, airlift will be inadequate and inappropriate for first-week deployment of even light mechanized units (although some heavy equipment can always be deployed if needed badly enough). Instead, airlift should be focused on light mobile infantry in "brigade-sized" task forces. An important factor here is that airlift would be needed for many other purposes simultaneously (e.g., to deploy support for tactical air forces and missile defenses, and to provide supplies for humanitarian relief).

- Although technology could greatly improve the sailing speed of sealift, a limiting factor would be the times required for loading, en route refueling, unloading, assembly, and initial maneuver. As a result, we do not see rapid-sealift (or airship) options providing first-week capability—even if they prove attractive and affordable for the long run.

These conclusions are rather negative. What *is* feasible, in contrast, is to use ship-based prepositioning to deliver light mechanized (and heavier) forces in a week—assuming use of strategic warning for ships not already in the region.

Recommendation

- We recommend that the Department of Defense "zero base" (i.e., rethink without being biased by current practices) its use of current Marine Corps Maritime Prepositioning Force Squadrons and Army Afloat Prepositioning Sets with the objective of providing first-week light or medium-weight mechanized forces that could—with modern doctrinal concepts—be aggressively employed immediately, rather than fighting in more classical and deliberate ways. With such rethinking, initial versions of such forces could be operational within five years. Major enhancements could then occur in subsequent years, building on experience.

- Consistent with this, the Department of Defense should more routinely move prepositioning ships upon warning—as carrier battle groups have for decades. This is largely a high-policy issue. Marine Corps experience in the Persian Gulf shows that high readiness can be maintained with on-station ships.

Finally, because the threat of mass-destruction weapons may preclude using major forward ports, "over-the-shore" operations and replenishment from the sea—or deployments into distant ports followed by long-distance maneuver—will sometimes be necessary. Such operations, which would be slower, will require changes in equipment and doctrine, and will probably be easier with lighter mechanized forces that have few or no 60–70 ton tanks and a much reduced logistics burden.

A RECOMMENDED OPERATIONAL CONCEPT

Given that the United States could—with current strategic mobility assets used differently than today—deploy light mobile infantry forces within about three to four days and light (or medium-weight) mechanized forces within a week, what operational concept and force structure would make sense if early actions are essential? In developing our answer, we avoided assumptions about whether Army forces, Marine Corps forces, or both would be used. We did so because we see opportunities for healthy competition and because, in practice, a rapidly deployable joint task force would often need units from both services.

Our recommendation is a proposed force with three components as follows (the detailed sizes and configurations are illustrative):

- An Early Allied-Support Force with a few hundred personnel that could in crisis greatly leverage the potential of allied forces already in place, primarily by linking to U.S. command and control, information systems, and long-range fires. This force may need to bring with it significant equipment for reconnaissance, surveillance, and communications.

- A Light Mobile-Infantry Force with 3,000–5,000 personnel organized into two principal types of units. The first would be 500-person light, mobile, air-deliverable fighting units with multiple missions such as defending critical facilities and launching missile attacks. The second would be infantry units with 50–80 personnel each that would operate forward (in some cases behind enemy lines) and—through networking—be able to direct long-range fires and conduct ambush operations.

- A Light (or Medium-Weight) Mechanized Force with 3,000–5,000 personnel in five or six agile tactical units capable, for example, of anti-armor missions against enemy forces already weakened by long-range fires and ambushes. It would draw on Air Force and Navy long-range precision fires, but would have some of its own (i.e., "organic") long-range missiles, shorter-range indirect fires, line-of-sight weapons, and attack helicopters. The light mechanized force concept (which some would call a medium-weight force)[1] would be a networked, highly mobile force with about five to six agile tactical units with organic lethality and survivability against enemy armor to complement early available standoff fires from land, sea, or air (and the contributions of allied ground forces). Ideally, major portions of the force would be fully air-assault mobile using C-130 or C-130 follow-on aircraft, but some larger equipment (even some tanks) may be needed. The force would deploy by fast sealift or maritime prepositioning (our preferred option), although some of its smaller and lighter units might be airlifted. Its systems would be designed for long endurance and economical expenditure of onboard consumables. Additional forces would, of course, reinforce as quickly as possible—arriving over a period of a month or two, primarily on existing fast sealift and amphibious shipping.

ENABLING TECHNOLOGIES

Much could be accomplished within the next five years with systems already available or in advanced development, but not adequately programmed. Doing so, however, would require elevating the priority of some programs. Candidates include operator-in-the-loop indirect-fire systems such as EFOG-M and loitering, short-time-of-

---

[1]The phrase "light mechanized" is ambiguous. It may mean fewer heavy items of equipment, lighter pieces of equipment, or both (our meaning).

flight systems such as LOCAAS. Major changes of doctrine will also be needed—e.g., the willingness to employ task forces with missile-firing units that would depend significantly for survival on allied forces, or with small teams that would operate forward (perhaps even in the enemy's rear area). In the longer run, achieving the full potential of the operational concept will require doctrinal and technological advances well beyond those now in hand. On the technology front, advances are especially needed in the following key areas: situational awareness, protection and armament, command and control, intra-theater lift, and vehicles for both transportation and fighting. In the body of the monograph, we identify what we see as the enabling technologies.

WHAT CAN BE ACHIEVED?

RAND analyses conducted for the 1996 and 1998 DSB summer studies and subsequent projects shed considerable light on the potential value of a mix of long-range fires and maneuver forces. The principal conclusions are as follows.

Long-Range Fires Used Alone

The most important conclusions relevant to this study are that

- Modern long-range fires alone may indeed deter or thwart a mechanized invasion in some circumstances, such as when the invader must cross essentially open terrain and maneuver long distances while the United States controls the air.

- However, their effectiveness in other circumstances may be drastically reduced as the result of terrain masking; enemy dispersal during maneuver; and long-range, mobile, air-defense threats that delay airborne surveillance and bombing by nonstealthy aircraft.

A number of factors underlie these sobering conclusions. First, small-footprint munitions carried by missiles launched from stand-off ranges (e.g., the sensor-fused weapons [SFWs] planned to be carried by the current JSOW) lose effectiveness against even moderately dispersed forces in open terrain because timing errors become important and programmed systems do not allow for en route updating. Second, effectiveness of all the long-range fires drops dramatically when the attacking force is moderately or substantially dispersed if that force is operating in mixed terrain with roads that appear canopied when viewed from long distances from aircraft. Moreover, in mixed terrain with hiding sites, the invader may use "dash tactics" in which his army is visible and vulnerable only a fraction of the day, thereby greatly reducing the effectiveness of aircraft flying sorties uniformly over time. As discussed in the text, all of these factors have large effects.

To be sure, technological responses are possible for all of these problems. In particular:

- Larger-footprint weapons such as BAT (Brilliant Anti-Armor munition) and LOCAAS (Low-Cost Anti-Armor System) are effective against moderately dispersed forces in the open.

- Standoff weapons could be given en route updates on projected target locations, which could reflect post-launch tracking and detailed computerized knowledge of the roads and terrain.

- Capabilities for suppressing enemy air defenses (SEAD) can be improved, which would allow earlier use of direct-attack aircraft weapons (although *assuring* rapid suppression may be impossible in some cases, as Kosovo demonstrated).

- Satellites or reasonably stealthy unpiloted aerial vehicles could detect and track vehicles from the outset of war, without waiting for SEAD as vulnerable aircraft such as J-STARS may need to do. Space-based moving-target-indicator (MTI) radars could also greatly improve the targeting of vehicles moving in mostly canopied mixed terrain.

- Ground or naval missiles could be used before SEAD was complete.

- Short-time-of-flight weapons—whether from loitering aircraft, hypersonic missiles, or shorter-range unmanned "missiles in a box"—could be effective and efficient against dispersed forces in canopied terrain, especially if detection and tracking systems had significant foliage-penetration capability. They would also be much more effective than nonloitering aircraft against armies using "dash tactics."

Unfortunately, although such technological fixes could improve the effectiveness of long-range fires, it may be many years before such fixes are introduced and the necessary inventories procured—especially since these improvements are nearly all expensive. Further, other countermeasures to long-range fires exist. Thus, despite the great value and potential lethality of long-range fires, we conclude that it would be unwise to *assume* that long-range fires acting alone will be sufficient in complex terrain or in the absence of defensive depth.

Combined Use of Fires and Ground Forces

All of this suggests combining long-range fires and maneuver forces. Our conclusions from analysis to date are as follows:

- An early deploying ground component should be viewed in task-force terms because it would likely have a variety of traditional infantry, missile-firing, or attack-helicopter missions in addition to providing eyes on the ground and independent ambush capability.

- Early employment of ground-based long-range missiles carrying large-footprint munitions such as the Army's Tactical Missile System (ATACMS) with BAT munitions could in some cases be especially useful if suppression of air defenses lagged and made direct-attack operations by tactical aircraft difficult. Similarly, even small attack-helicopter forces could be powerful early against fast-moving, exposed enemy forces (the helicopters should not be expected to make deep attacks through defended zones).

- Even relatively small maneuver forces could materially affect the invader's maneuver tactics—forcing greater concentration and more deliberate movements. Both effects could dramatically enhance the effectiveness of long-range fires, as evidenced by recent events in Kosovo.

- If small maneuver units can be inserted behind the lines early—or inserted where enemy forces pass by them—primarily to improve targeting for long-range fires, they should not depend solely on those long-range fires for their own survival and effectiveness. When random factors and intelligent enemy tactics are considered, uncertainty analysis suggests that the risks to such units would be quite high unless they had both tactical mobility (e.g., from advanced two- to three-ton vehicles) and significant organic indirect precision-fire capability (e.g., with the EFOG-M and HIMARS weapons systems, which operate over a range of about 30 km). Such units, however, would not need—and would be burdened by—traditional weapons such as 155-mm artillery.

With the caveat of the last bullet, agile, dispersed insertion forces appear attractive for many circumstances of interest. In human-moderated simulation, their distributed operation and tactical mobility allowed a high level of survivability and lethality—*assuming U.S. information dominance, which would be essential to their survival*. In particular, they were quite successful in ambushing enemy forward combat armored units and combat-service support vehicles. Moreover:

•   Given organic indirect precision fires, the combination of such forces and long-range precision fires was much more effective than long-range fires alone.

Also, when—instead of ambushing enemy combat forces—these simulated forces attacked support elements in the enemy's rear area (e.g., resupply vehicles, C2 centers, air defense sites, assembly areas, and artillery units), they suffered an order of magnitude fewer losses. Experienced officers observing the simulation concluded subjectively that such attacks on support structure might well have blunted the momentum of the combat units.

Command, Control, Communications, Computers, Intelligence, Surveillance, and Reconnaissance (C4ISR)

Many aspects of our analysis involved alternative sets of C4ISR capabilities. In our human-in-the-loop high-resolution analysis we observed that a comprehensive intelligence preparation of the battlefield (IPB) and an effective reconnaissance and surveillance plan are important for identifying good insertion locations, entry routes, and exit routes for the small, agile, light mobile-infantry units. Thus,

C4ISR supporting these activities matters a great deal. In contrast, some other C4ISR improvements were less significant than expected. In particular, having perfect rather than approximate information on mobile targets at the time weapons were launched from standoff ranges had only modest effects in canopied terrain because results were limited by errors in projecting future target locations, and by effects of terrain on both targeting and munitions effectiveness. However, providing accurate and timely target location information to the forward operating ground forces improved their lethality significantly. Overall, then, improved C4ISR matters, but not always as one might expect. Equally important is denying the enemy surveillance and reconnaissance: Information dominance requires this *and* good C4ISR. Achieving it will often be difficult, even in the future.

MEASURING THE POTENTIAL VALUE OF EARLY GROUND FORCES IN THE HALT PHASE

The value of our postulated early ground forces would, of course, be highly scenario dependent. Consequently, we constructed a vast test set of scenarios by considering all the combinations of various assumptions about the size of the threat, its base movement rate, the number of axes of advance, the time to suppress air defenses, deployment rates, strategic warning, and tactical warning. We also varied—for cases that included the advanced ground force—assumptions about how much attrition it could cause by ambushes and leveraging allied capabilities, and how much effect it and allies could have in reducing the enemy's rate of advance. The resulting test set is just that, a test set. Nonetheless, it is useful in evaluating the potential benefit of the advanced ground force.

Because of its familiarity, we again used the overall scenario of Iraq invading Kuwait, but this time attempting to go into Saudi Arabia. We could equally well have chosen a number of alternatives. Indeed, doing so might have been more favorable to ground-force use because, in mixed terrain, such teams might be more able to find suitable ambush sites. Nonetheless—as suggested not only in simulation but also in the Marine Corps Hunter Warrior experiments and the Army's experience at the National Test Center—so long as the United States could deny Iraq use of the air for surveillance, many such ambush sites would exist and ground forces could plausibly operate and survive.

Table S.1 shows results for the Persian Gulf scenarios. It is based on using Monte Carlo techniques to select 300 cases from the overall space of test-set cases. Table S.1 displays results for 0, 5, and 10 days of tactical warning, and for cases with fires only and with both fires and maneuver forces. To avoid excessive quantification of inherently uncertain simulations, we show the general extent of Red's maximum advance in terms such as "key oil facilities." We show results for median outcomes and, in adjacent columns, we show the range of maximum advances covering about 75 percent of the cases. We see that the combined use of long-range fires and maneuver forces could have a substantial effect strategically—holding penetrations to

Table S.1

Potential Value of a Rapidly Deployable Ground Force for Defense of Kuwait and Saudi Arabia
(outcomes across test set of scenarios)

	Deepest Iraqi Penetration (median from test set of scenarios, followed by a range covering 75% of cases)			
	Without New Ground Component		With New First-Week Ground Component	
Tactical Warning (Days)	Median Outcome	Range of Outcomes	Median Outcome	Range of Outcomes
0	Key oil facilities (~500 km)	Gulf coast to Dhahran (~350–580 km)	Gulf coast (~400 km)	Northern Saudi Arabia to key oil facilities (~220–550 km)
5	Gulf coast (~480 km)	Northern Saudi Arabia to Dhahran (~320–580 km)	Northern Saudi Arabia (~200 km)	Kuwait to Gulf coast (~100–380 km)
10	Gulf coast (~400 km)	Northern Saudi Arabia to Dhahran (~200–550 km)	Northern Saudi Arabia (~100 km)	Kuwait to northern Saudi Arabia (~75–350 km)

NOTES: Test set assumes the threat of mass-destruction weapons, which causes reduced sortie rates and deployment rates. The set assumes a threat greater than today's but varies size of threat, base movement rate, number of axes of advance, SEAD time, deployment rates, strategic warning, and tactical warning. For the light mobile-infantry case, it also varies when that force is available, how much attrition it can cause, and how much effect it (and allies) have in reducing the rate of advance.

Kuwait or Northern Saudi Arabia in many cases, and reducing the likelihood of penetrations as far as the Gulf coast, much less to the principal oil facilities. Although the analysis is only illustrative, it demonstrates that one can quantify the potential benefit of the proposed forces—so long as one considers a range of situations rather than focusing only on some approved baseline, such as a scenario that assumes a week or so of actionable tactical warning.

CONCLUSIONS

A joint-task-force approach combining long-range fires with a rapid deployment ground force consisting of an Early Allied-Support Force, a Light Mobile-Infantry Force, and a Light Mechanized Force has a great deal of potential as the front end of a larger campaign. This would be true for major wars and—as discussed in the main text—for certain kinds of small-scale contingencies. Indeed, the potential is so great that we recommend vigorous efforts, including service and joint experiments, to establish a more reliable empirical base. These should include more stressful field experiments to characterize and assess: the leverage achievable by connecting defended allies to U.S. C4ISR systems; the effectiveness and survivability of small teams used at the front or in the enemy's rear area; and the nuts and bolts of inserting, extracting, and supporting such operations with C4ISR, fires, and logistics.

It is possible that a sober empirical assessment will conclude that some aspects of the ground-force concept—primarily the deep insertion of small teams for ambushing—will be too risky for commanders to embrace. But the opposite conclusion is also possible—at least for instances in which the United States is able to prevent enemy surveillance and has good intelligence on the battlefield. Moreover, other aspects of the concept (e.g., the value of inserting unmanned "missiles-in-a-box" systems and the value of deploying long-range missile batteries, attack helicopters, and mobile infantry early) may prove more robust. In any case, vigorous research, experimentation, and analysis are badly needed.

Major questions still exist regarding some aspects of the concept, but we believe that other aspects are well enough understood that the United States should seek initial versions of an operational first-week ground-force capability within five years. This will require changes in the use of strategic mobility, doctrine, and program priorities. A top priority should be "zero basing" the use of current ship-based pre-positioning assets (and airlift) to enable a near- to mid-term version of the advanced joint task force. Initial forces would be heavier and less capable than technology will make possible in the longer term, but much can be done within the five years. Establishing such a near to mid-term goal would be liberating to military innovators, who have been hampered by an excessive emphasis on the distant technology of super-light forces and advanced lift. Moreover, such a rapid effort would be an excellent way to invigorate DoD's effort to "transform U.S. forces" for the needs of future decades.

ACKNOWLEDGMENTS

We appreciate our collaboration with Brigadier General Huba Wass de Czege (USA, retired) during the 1998 Defense Science Board summer study. He contributed heavily to our initial concept for the ground-component task force, although its final version in this monograph has evolved substantially and is our responsibility. General David Maddox (USA, retired) and Lieutenant General Paul Van Riper (USMC, retired) also provided helpful comments during and after the DSB study. Major Steve Struckel (USA), Lieutenant Colonel Jay Bruder (USMC), and Lieutenant Colonel Phillip Schlatter (USA)—military fellows posted at RAND—had many helpful comments and suggestions on earlier drafts. Colleagues Jed Peters and Tom McNaugher provided thoughtful formal reviews.

ACRONYMS

Although we have attempted to make the text readable without prior knowledge of the bewildering alphabet soup of acronyms, we have also used the common acronyms so that readers can connect our discussion with the concepts and systems discussed elsewhere.

Acronym	Definition	Comment
AAN	Army After Next	Conceptual force planned for 25–30 years in future
ACV	Advanced Combat Vehicle	
AFSAB	Air Force Scientific Advisory Board	
AFSS	Advanced Fire Support System	Example: short-range PGMs with automated "pods" air dropped (also called "missiles in a box")
AFV	Armored Fighting Vehicle	Examples: Bradley fighting vehicle and Russian BMP
AGS	Armored Protective Gun System	Light (18–24 ton) tank
AHMV	Advanced High-Mobility Vehicle	Small (2 to 2–1/2 ton) wheeled vehicle
AOE	Army of Excellence	Current force configuration
APC	Armored Personnel Carrier	Examples: Bradley fighting vehicle and Russian BMP/BRDM
APS	Advanced Protection System	Example: short-range anti-missile missiles such as those developed by Russia

ARES	Advanced Robotic Engagement System	
ASE	Advanced Aircraft Survivability Equipment	Electronic jamming and spoofing package
ATACMS	Army Tactical Missile System	Long-range missile fired from HIMARS or MLRS launcher
AWE	Advanced Warfighting Experiment	
BAT	Brilliant Anti-Armor munition	Carried on, e.g., ATACMS
BLOS	Beyond Line of Sight	
BLT	Battalion Landing Team	
C2	Command and Control	Decisionmaking functions
C4ISR	Command, Control, Communications, Computers, Intelligence, Surveillance, and Reconnaissance	Adds intelligence collection, dissemination, presentation. In practice, usually does not refer to higher-level C2.
CEC	Cooperative Engagement Concept	
DARPA	Defense Advanced Research Projects Agency	Focuses on funding innovative research
DAWMS	Deep Attack Weapons Mix Study	A recent DoD examination of options
DoD	Department of Defense	
DRB	Division Ready Brigade	The brigade of the 82nd Airborne Division that is at top readiness (this rotates among brigades)
DSB	Defense Science Board	
EFOG-M	Extended-Range Fiber-Optic Guided Missile	A precision anti-tank weapon with a beyond line-of-sight range of perhaps 20 km. Can exploit a separate "spotter" located within line of sight to the target.
EM/ET gun	Electromagnetic/ Electrothermal gun	"Railgun" powered by electrical energy (augmented by chemical energy in ET version)
FCV	Future Combat Vehicle	Conceptual 20–40 combat vehicle

FLIR	Forward-Looking Infrared	The principal type of sensor used for infrared devices in tanks, aircraft, and other weapons systems
FOTT	Follow-on to TOW	Advanced short-range, direct-fire anti-armor missile
FRV	Future Reconnaissance Vehicle	A postulated semi-stealthy high-mobility vehicle that would replace the M-3/HMMWV reconnaissance platforms
HEAT	High-Energy Anti-Tank	A shaped-charge chemical-energy warhead
HIMARS	High-Mobility Artillery-Rocket System	A smaller, wheeled version of the current MLRS
HMMWV	High-Mobility Multipurpose Wheeled Vehicle	4–5 ton successor to venerable jeep
IFV	Infantry Fighting Vehicle	
IPB	Intelligent Preparation of the Battlefield	
IRC	Immediate Ready Company	
JSOW	Joint Standoff Weapon	Munition dispenser lofted from aircraft
J-STARS	Joint Surveillance Target Attack Radar System	An airborne system used since the Gulf War, especially for detecting and tracking moving vehicles
JTF	Joint Task Force	Command echelon controlling joint assets
JTR	Joint Tactical Rotorcraft	Tactical airlifter with ~10 ton payload
KEP	Kinetic Energy Penetrator	Nonexploding warhead, typically a long rod penetrator
KLA	Kosovo Liberation Army	
LAV	Light Armored Vehicle	16-ton Marine Corps wheeled vehicle
LOCAAS	Low-Cost Anti-Armor System	Loitering munition with roughly 30 minutes of endurance

LOSAT	Line-of-Sight Anti-Tank	Hypervelocity (~Mach 7) direct-fire missile
LRF	Long-Range Precision Fires	
MAGTF	Marine Air-Ground Task Force	Can be of highly variable size and composition depending on tasks at hand. Might be of MEU or MEB size.
MEB	Marine Expeditionary Brigade	Brigade-sized unit that may employ amphibious ships or maritime prepositioning ships
METT-T	Mission, Enemy, Terrain and Weather, Time, and Troops Available	Considerations used by officers in developing plans
MEU	Marine Expeditionary Unit	Regimental-sized afloat unit
MLRS	Multiple Launcher Rocket System	Launcher of ATACMS
MPFS	Marine Prepositioning Force Squadron	
MPS	Maritime Prepositioning Squadrons	
MTW	Major Theater War	High-intensity, large-scale conflict
NTC	National Test Center	
OCSW	Objective Crew-Served Weapon	Advanced machine gun
OSD	Office of Secretary of Defense	
PGMMs	Precision Guided Mortar Munitions	Weapons able to detect and hit specific targets or coordinates
PPI	Preplanned Product Improvement	
PSC	Precision-Strike Company	Advanced unit concept
PST	Precision-Strike Team	Advanced unit concept
PSYOPS	Psychological Operations	
RFPI	Rapid Force Projection Initiative	A research program that preceded the Army After Next effort
RMA	Revolution in Military Affairs	
RSTA	Reconnaissance, Surveillance, and Target Acquisition	Sensor functions

SADARM	Sense and Destroy Armor	Smart munition for 155-mm artillery. Primarily for attacking light armor.
SEAD	Suppression of Enemy Air Defenses	
SFW	Sensor-Fused Weapon	
SLID	Small Low-Cost Interceptor Device	Active protection system for ground vehicles
SOF	Special Operations Forces	
SPH	Self-Propelled Howitzer	Examples: U.S. 155-mm Crusader and Russian 152-mm 2S3
SSC	Small-Scale Contingency	Any contingency short of an MTW
SSTOL	"Super" STOL	A postulated follow-on to the C-130. Intended for use from medium-sized carriers.
STOL	Short Takeoff and Landing	Requires less than 1,000 feet under full load
SUO	Small Unit Operations	A DARPA-postulated RMA force
SWA	Southwest Asia	
TACMS	Tactical Missile System	
TOF	Time of Flight	
TRADOC	Training and Doctrine Command	
UAV	Unmanned Aerial Vehicle	Examples: Predator (in use) and Global Hawk
UCAV	Unmanned Combat Aerial Vehicle	An armed UAV
WCMD	Wind-Corrected Munitions Dispenser	
WMD	Weapons of Mass Destruction	Chemical, biological, and nuclear weapons

Chapter One

INTRODUCTION

BACKGROUND

Challenges and Missions for Ground Forces

As the United States moves into the 21st century, it faces an uncertain and complex international environment. Given the nation's worldwide interests and responsibilities, the need for U.S. and allied military intervention in crises will undoubtedly continue. At the same time, uncertainties about threat and scenario details will require highly flexible military capabilities and—we believe—the ability for substantial action within *hours and days* rather than many days and even weeks.[1] If this is correct, then a high priority for the Department of Defense (DoD) is developing an early-intervention force, a task that can begin in the near term but will probably take many years to complete. Much discussion to date has focused on what could be accomplished with rapidly deployable long-range fires from aircraft and missile ships.[2] Strategically, however, it is clear that many challenges would require early use of *ground forces*—ground forces that could be deployed from the United States and effectively employed within days. Such a force would be quite different from the

---

[1]A separate issue is assuring capability for intervention when the United States does *not* act quickly as discussed in this monograph, but rather may decide to act only a week or so after hostilities begin. The capabilities discussed in this monograph would be useful in that instance also.

[2]See, e.g., Bowie et al. (1993), Johnson and Libicki (1995), DoD (1996), DSB (1996a,b), AFSAB (1995), Davis et al. (forthcoming), Davis and Carrillo (1997), NRC (1997a), DSB (1998a,b), and Davis, Bigelow, and McEver (1999).

Marine Corps and Army forces feasible in the near to mid term, but it would build on innovations that the two services are already pursuing.

Such rapidly deployable ground forces could have many missions, not just missions associated with clear-cut major theater wars such as the 1990–1991 Gulf War. These include—for enemies both large and small, and in theaters with open, mixed, and rough terrain (including urban sprawl) where long-range fires have limitations[3]—

- *Deterring an invasion* of a friendly nation by reinforcing its forces convincingly during crisis without necessarily deploying a massive force or taking other actions that are difficult to reverse or repeat.

- *Halting an invading army* when a regional power attacks a friendly nation with short actionable warning and the friendly nation lacks the depth that might permit long-range aircraft and missiles to carry the burden of defense as ground forces deploy.

- *Deterring or preventing imminent "ethnic cleansing"* of a minority population, as occurred most recently with Serbian actions within Kosovo, which NATO was unable to defeat quickly for lack of forces on the ground where the atrocities were being committed.[4]

- *Destroying weapons of mass destruction* (WMD) that an enemy is about to employ or that the enemy is using to coerce the United States or its allies—including deeply buried mass-destruction weapons that cannot be neutralized from the air.

---

[3]Some of the likely difficulties are discussed as Achilles' heel problems or asymmetric strategies in, e.g., Davis, Gompert, and Kugler (1996), Bennett (1995), Cohen (1999), NDP (1997), and Bennett, Twomey, and Treverton (1999). John Foster led an influential Defense Science Board study on such matters in 1995. For a discussion oriented toward Marine Corps missions, see USMC (1998a).

[4]Airpower won its first war in Kosovo, but NATO failed in its first objective of preventing ethnic cleansing, which could not have been prevented without rapid-action ground forces (except in the event that initial bombing deterred such actions). Whether the risks associated with using rapid-action ground forces would be acceptable would be highly situation dependent. Arguably, however, the United States needs the option.

- *Compelling an invader to reverse course* and withdraw by stabilizing the theater and rapidly constructing the capability to counterattack and to threaten the invader's army and homeland.

To be sure, if the United States had the capabilities needed for these missions, then deterrence might often be achieved without crisis or, if crisis occurred, without the need actually to deploy and employ the forces. In this monograph, however, we discuss matters as if the missions are actually being conducted.

Approach of the Study

Against this background, the present monograph is a think piece describing one vision of a rapidly deployable early-entry ground force suitable to an enduring U.S. strategic challenge: halting, deterring, or blunting enemy invasions of friends and allies during the first days and week of an invasion. Our vision is unabashedly ambitious and technologically optimistic because it is driven by our perception of future operational needs. In the remainder of this introductory chapter we present an overview of the issues involved. Chapter Two describes our operational concept in more detail and sketches an image of unit configurations, equipment, and systems for a suitable force. Our purpose is certainly not to define such a force definitively, but to provide a framework that can help guide planning on many issues. Such a framework can be quite helpful even if details of the illustrative image are flawed. In particular, achieving the postulated capabilities will require innovations and technological advances well beyond currently emerging precision weapons and planned networking. Thus, in Chapter Three we identify enabling technologies and note doctrinal developments and research (including field experiments) that seem particularly crucial. In Chapter Four, we present some initial analysis of what the future forces we envision might accomplish. This analysis is actually rooted in midterm, within-reach systems and concepts, and can therefore be relatively concrete. It suggests that much could be accomplished over the course of the next three to seven years, with emerging technologies providing ever-improving capabilities over time. Chapter Five provides some conclusions and cautions.

OVERVIEW OF IMPORTANT FACTORS

Tactical Considerations

Many reasons for a greatly improved early-entry ground-force capability relate to details of the tactical situations that might be faced by future task force commanders. For example, the United States must anticipate that many aggressors could quickly overpower the ground forces of a friendly nation if those forces were left to themselves. Historically, we know that the collapse of such defenses could occur within the first few days. It would therefore be important to strengthen the potential of allied forces and to optimize the use of first-hour combat power.

A second reason for early ground-force action is that an aggressor will often want to break up the ally's forces and intermingle his forces with theirs, thus complicating targeting and raising risks of fratricide. The attacking enemy will also want to rapidly secure urban centers for political and operational reasons. If an aggressor succeeded in such actions, precision weapons would be less effective and the United States would have lost a significant part of its technological advantage.[5]

A third reason is rooted in issues at the level of individuals and individual systems on the complex battlefield. Even if we consider the cases usually thought of as ideal for "immaculate warfare" fought with long-range fires launched from standoff range, deeper analysis by RAND and others has consistently uncovered profound problems with such cases that are unlikely to be resolved for many years. In some of these cases, the problems may never be resolved. As we demonstrate analytically in Chapter Three, the task for long-range fires acting alone becomes increasingly difficult as the invader takes sensible tactical and technical countermeasures, and when he is able to exploit natural and man-made terrain.[6] However, the

---

[5]Such actions will occur despite best efforts to pursue ideas such as those in this monograph. It is therefore necessary for the Marine Corps and the Army to maintain and improve capabilities for "dirty" operation, such as those in urban sprawl, but these operations are not our focus here.

[6]Analysts divide on these matters. By and large, technologists tend first to think of plausible future capabilities that could counter the counters that can be identified

combination of long-range fires and ground-force maneuver units appears to be much more powerful than either fires or ground forces alone. They can both reinforce each other and hedge against each other's failures—which cannot be reliably predicted.

With this background, then, let us next attempt to characterize the capabilities the U.S. needs for early intervention, assess what is feasible with respect to strategic mobility, and infer from that the kind of operational concept that might make sense.

Capabilities Needed

Timeliness in Major Theater of War Interventions. When considering future major theater wars (MTWs) and some important classes of small-scale contingencies (SSCs), it quickly becomes evident that the United States needs the ability to intervene effectively in days, not weeks.[7] This has profound implications. Let us illustrate this first for MTWs.

The reason for wanting to deploy troops within a few days in some MTWs is the short distance an invading force typically needs to go to achieve an initial set of objectives and become entrenched. For example, the Iraqis are less than 100 km from Kuwait City. Most important oil fields are within 400 km of Iraq's border. Rapid reaction forces that take a week to arrive might then have to eject a well-defended opponent in cities.

A great deal of analysis has been done on the "halt problem" (halting the advance of an armored invasion force, so that forces can then be

---

against current and projected U.S. capabilities. However, at any given time there is typically a many-year gap between the capabilities that can or *may* be achievable technically and what in fact exists. Similarly, there tends to be a large gap between what studies assume in the way of force-employment optimality and what seems to occur in the real world. As merely one example, "unmanned combat aerial vehicles" (UCAVS) may have a glorious future, and might be able to accomplish a great deal without friendly personnel on the battleground itself, but they do not exist today and, even if they did, would likely have serious shortcomings for many years. In this monograph, we attempt simultaneously to be bullish about technology, but sober about what it can truly accomplish in different time frames and circumstances.

[7]Other authors discuss overlapping concepts with terms such as rapid decisive operations, rapid dominance (Ullman and Wade, 1996), early-entry forces, and early-intervention forces.

built up for a subsequent counteroffensive to destroy the enemy's army and restore territory). That analysis shows that achieving an *early* halt is very difficult unless the United States has substantial forces already in place along with allies (e.g., as in Korea). Results depend on many factors[8] as discussed later in the monograph, but the basic problem is depicted in Figure 1.1. It shows, notionally, outcomes of efforts to halt and defeat an enemy early in different regions of a simplified "scenario space." This display is a bit different from more usual ones. The horizontal (x) axis shows a timeline dimension of the scenario: when full-scale deployment begins, relative to D-Day. The vertical (y) axis measures the difficulty of the scenario in terms other than time. Thus, "up" corresponds to a larger threat, a more competent threat, clever enemy strategies, use of mass-destruction weapons, and so on. Results are indicated not by one of the axes, but by color or shading.[9] In this depiction, worst cases are at the top right corner (i.e., large threats adopting strategies effective against U.S. and allied weaknesses); the best cases (minimal threat, early deployment) are at the bottom left.

Based on extensive analysis that accounts for many detailed but highly uncertain variables, we characterize U.S. capabilities as shown in the top panel: With current and programmed forces, success would be likely only in the bottom left (light) region. In contrast, if the United States had a rapidly deployable joint task force of the kind we envision, then—as indicated in the lower panel, and as discussed further in Chapter Four—the envelope of success could be pushed upward and to the right (the shaded region). Some cases, of course, would still be too difficult (those in the top right). The solid dot in the success-with-current-forces region suggests that, for common

---

[8]See Davis and Carrillo (1997), Ochmanek et al. (1998), and Davis, Bigelow, and McEver (1999). Prospects are improved when the terrain is open, air forces already exist in the region, and a "leading-edge attack" is feasible against an attacker moving on only one or a few axes of advance.

[9]This is a particular example of a "scenario-space" diagram. It has the advantage of displaying capability over quite a wide range of assumptions, instead of for a specific scenario. This, in turn, encourages thinking of capability improvements as initiatives that would increase the fraction of the scenario space for which capabilities would be adequate. This is quite different, obviously, from measuring an initiative's value for its ability to improve results somewhat in a highly specific scenario. In Chapter Four, we present a similar figure based on a specific test set of scenarios.

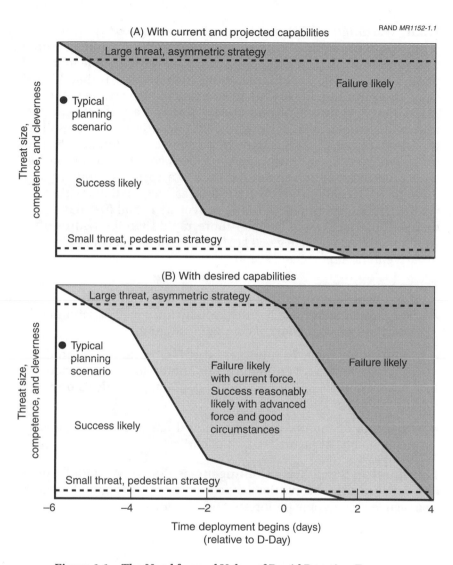

Figure 1.1—The Need for and Value of Rapid Reaction Forces

planning scenarios, outcome prospects look good even with current and programmed forces and sizable threats—but only because considerable actionable warning time is assumed. That assumption is simply not prudent.

What "Warning Time" Can Reasonably Be Assumed? What warning time *should* we assume? Here we depart from the assumptions often made in studies that contemplate actionable warning time of a week or so. One rationale used for assuming such warning is that, historically, such warning in fact has been available in most cases: crises seldom emerge overnight and enemy forces are seldom ready for immediate employment.[10] However, the incentives are now so high for "short-warning" actions that we must expect potential adversaries to reduce *actionable* warning time. This could be accomplished with prepositioned forward logistics and a variety of military and diplomatic deceptions involving exercises, negotiations with pleas for the United States to do nothing provocative, and first-day movements—once aggression begins—more rapid than the figures often cited for advance rates (e.g., 60 km/day).[11] The targets might be, e.g., urban centers, airports, and seaports.

The conclusion, then, is

- The United States needs substantially improved capabilities for intervention within *days* of combat's initiation.

- In planning such capabilities, the United States should *not* assume "tactical warning" (a misnomer for what is actually "usable deployment time prior to conflict, given decisionmaker uncertainties").

- As a rule of thumb, planning should assume that large-scale deployments begin on D-Day, not earlier.

Timeliness in Small-Scale Contingencies (SSCs). Without elaboration, let us merely observe that something similar to Figure 1.1 would also be relevant in some important SSC missions.

---

[10]A considerable literature exists regarding strategic warning in historical crises. That literature is generally depressing because so many examples exist where strategic warning was available, but ambiguous, and preparations for defense came much too late. See Knorr and Morgan (1983), Betts (1982), and, in the context of conventional arms control, Davis (1988).

[11]Historically, the traditional *daily* advance has been only about what modern units can move in an hour or so. Advance rates have been limited by doctrine, logistics, and tactical-level command-and-control difficulties in efforts to maintain a disciplined march formation. This means that the *potential* for much faster movements exists.

1. *Ethnic Cleansing*. Because of the short distances typically involved, and the ability of infantry or paramilitary forces to operate without creating large and highly visible targets, those intent on ethnic cleansing activities cannot realistically be thwarted without intervening ground forces. And, since murders, rapes, and the torching of cities can be done within a few days or a week, it follows that a prevention force—*if* it could be employed with acceptable risk—would need to begin operations on that same short timescale or faster. How much force would be needed would depend on the size of the area involved, the size and character of the ethnic cleansers, the quality of intelligence, and the size of any friendly defender forces. Something like Figure 1.1 would apply. It should be noted that U.S. forces might not be able to stop enemy infantry movements and actions, but they could attack command structures, logistical nodes, supporting vehicles, and supplies. This could quickly cause havoc and cause the noxious activities to cease.

2. *Destroying Mass-Destruction Weapons.* If the United States were to consider preemptive destruction of mass-destruction weapons, it seems likely that the decision to attack would be made only very late in crisis (e.g., hours before large-scale hostilities began) or after war was under way, in which case the mass-destruction weapons might be used at any time. Again, speed would be essential. A small raid against a single facility might be accomplished with a special operations forces (SOF) team, but larger attacks against better-prepared defenders would quickly increase the size of the force needed. Something like Figure 1.1 would apply.

3. *Decisive Deterrent/Compellent Operations Without War.* If the United States were faced with imminent aggression that might still be deterrable, or if the aggression's first stage had occurred but further moves were imminent, then a sufficiently decisive show of force might be enough to salvage the situation. Naval and Air Force activities might or might not be sufficient initially, depending on the specifics of the situation and terrain involved. They were not effective against Slobodan Milosevic in the days prior to his moves against ethnic-Albanian Kosovars. If ground-force operations were needed, then something like Figure 1.1 would again apply, although the force levels needed quickly (for

potential employability) would be smaller than if war actually began.[12]

Some of this may seem obvious, but we are struck by the reported surprise of many NATO leaders when NATO's deterrent actions against Milosevic failed. An important and predictable reason for the failure is that both political and military leaders reach a point in their decisionmaking at which the "die is cast." This reflects a universal feature of human cognition and decisionmaking. Immediate deterrence and compellence often depend on acting during the window in time between the start of crisis and the mental "casting of the die."[13] Speediness in crisis action, then, is important for psychological as well as physical reasons.

Logistical Considerations. Another consideration in thinking about future forces is that some contingencies require deep and dispersed operations that make logistical resupply very difficult.[14] Logistics will also be complicated because the threat of weapons of mass destruc-

---

[12]It is noteworthy that the actions of the Kosovo Liberation Army (KLA) ground forces in Kosovo, during the last week or so of the conflict, reportedly had a large effect: the KLA's attacks, although small, forced the Serbians to concentrate forces and protect particular positions. This, in turn, made it easier for NATO surveillance to detect and air forces to attack them successfully. See Secretary Cohen's June 10, 1999, briefing on "Operation Allied Force," available on the web at www.defenselink.mil./news/Jun1999/g990610-J-0000K.html/.

[13]For further discussion of cognitive models of decisionmaking and their relationship to deterrence and compellence, see Appendix G in NRC (1997b), or Davis (1994). That material was drawn in part from a study of Saddam Hussein conducted in 1990–1991. The notion of a die-is-cast phenomenon is probably familiar to all of us from our own actions. But at the national level, one might think of President Bush's obvious disinclination to negotiate a late-in-the-game Iraqi withdrawal from Kuwait once the preparation for Desert Storm's counteroffensive was well advanced (Pape, 1996, Ch. 7). An even better example is probably President Truman's decision to go ahead with the A-bombing of Japan despite some late-in-the-game signals from Japan about a willingness to negotiate. See Cousins (1987, pp. 40–42). For a more nuanced discussion of the ambiguities of attitude within the Japanese government, see Pape (1996, p. 123).

[14]Even today's Marine Corps sometimes employs small expeditionary units over long distances through the use of helicopters. Their missions include evacuations and humanitarian actions. In the future, many combat operations will become long-distance operations because of the need for surface ships to remain well offshore (perhaps 100–300 km) because of threats from mines, patrol boats, and missiles. The "operational maneuver from the sea" considered fundamental to future Navy/Marine Corps doctrine must necessarily confront this issue. See National Research Council (1999) for discussion.

tion and large-area conventional munitions will make localized logistical buildups extremely risky. In practice, the same forces that we consider for their rapid deployability will also be more sustainable in such deep operations and will create fewer demands for large, localized supply areas.

WHAT IS FEASIBLE FOR IMPROVED STRATEGIC MOBILITY AND FORCE LIGHTENING?

What Needs to Be Lifted?

If time is so critical, what can be done to improve drastically the capabilities brought to bear in the first days and week of a conflict? The answer depends in part on what can plausibly be accomplished with improvements in airlift or sealift, or the lightening of forces. In this regard, we are a good deal more pessimistic than some authors. Our conclusion on this is due to our observing—after having seen or participated in many studies over the years—that

- The concept of a truly "light force" adequate to intervene in dangerous places is a false idol. A serious early-intervention force will need not only tactical mobility, but also protection from ubiquitous small and medium-sized weapons against which nothing short of armor (at least light armor) will be sufficient.

It is sometimes argued that a combination of stealth and mobility will provide the necessary survivability, but the argument fails if the forces have to maneuver through dangerous zones where hidden adversaries can operate from relatively short range. That, regrettably, will often be the norm, not the exception.

It follows, as elaborated in Chapter Two, that we see the need for an early-intervention force with a combination of light mobile infantry and light (or medium-weight) mechanized forces. With this in mind, let us now discuss what is feasible with strategic mobility. We consider both current ground forces and those that postulate a good deal of lightening.

Current and Projected Deployment Times

Today's ground forces are arguably ill-equipped and unsuited to accomplish rapid-action missions in shooting situations. They will become less capable with time as potential adversaries plan for short-warning operations, actual or threatened use of mass-destruction weapons, and exploitation of both natural terrain and urban sprawl where long-range fires have limitations.

Portions of the 82nd Airborne Division can be airlifted within days, but the 82nd lacks potency against a mechanized force in many circumstances and has very limited tactical mobility.[15] Although capable for many missions, it would typically not survive in attempting to halt an invading mechanized force. The 101st air-mobile/air-assault division lacks ground mobility, lacks potency if its attack helicopters must operate against extensive air defenses, and is currently not rapidly deployable because of the lift and logistical demands of its aviation. Today's heavy Army force is too bulky to be deployed quickly in a crisis—unless heavy stocks and equipment are already prepositioned—in which case brigade-sized units can be ready for action within a week or so. A regimental-sized Marine Expeditionary Unit or MEU is forward positioned for immediate action with Navy support,[16] but mounting Marine Corps operations at the brigade level (in Marine Expeditionary Brigades or MEBs) also requires a week or so if maritime prepositioning sets are used.[17] Larger operations, by either the Marine Corps or Army, would require weeks to prepare. Most important, even Army and Marine Corps prepositioned units are currently configured with the expectation of relatively classic operations with buildup areas, defense lines, concentrated forces, and so on. They are by no means light and agile.

---

[15]A good source for mobility data and analysis is Congressional Budget Office (1997).

[16]A MEU has a command element, a battalion landing team (BLT), an air combat element, and significant support. Overall, it is "regimental-sized." When a MEU can coordinate with the leading edge of a maritime prepositioned unit, a second BLT can be available very quickly.

[17]For afloat prepositioning, actual times depend on prior readiness, use of warning to move ships to the region, configuration of equipment on the ships, availability of suitable ports or preparation for unloading "over the shore," weather, and other factors.

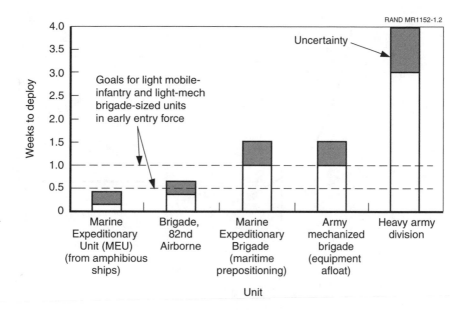

Figure 1.2—Deployment Times for Current Ground-Force Units

Figure 1.2 shows some of this schematically. Each bar graph indicates roughly (with shaded area for uncertainty) how long it takes to deploy a given unit. The horizontal lines suggest how quickly deployment may need to be accomplished (e.g., three to four days for a brigade-sized force, starting on the first day). Let us now discuss the issues for airlift and sealift separately.

What Is Feasible With Strategic Airlift and Lighter Forces? Figure 1.3 shows estimates of airlift time as a function of how much of our current and projected military airlift capacity is used (i.e., not counting use of the civilian reserve airfleet) and the weight of the units being moved. The effective airlift capacity (Cohen, 1999) may be a bit optimistic because of assumed efficiencies that have not been realized in real-world operations, but it is at least technologically and operationally feasible, given prior alert and the availability of bases. If, for example, we assume that 40 percent of capacity is used for the ground-maneuver forces we are concerned with in this report, then deploying the 82nd Airborne's division ready brigade (DRB) could

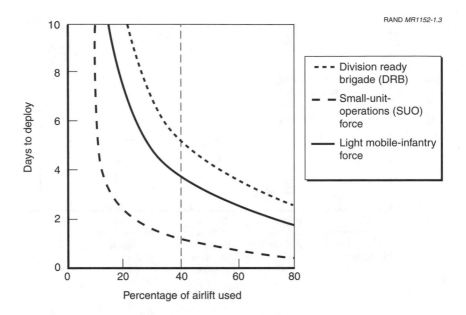

Figure 1.3—Airlift Times for Light Forces

take about five to six days, whereas deploying the super-light DARPA small-unit-operations (SUO) force would take only about one to two days. The force we describe in Chapters Two through Three would be closer to the DRB than the SUO, and is shown in the figure as requiring about four days.

Unfortunately, there would likely be many claimants for airlift during a crisis. Tactical air forces require substantial support equipment to be airlifted to the bases they will be using. The extent of this can be reduced, but not so much as might be expected naively, because of chronic reluctance to preposition certain top-of-the-line systems, equipment, and weapons abroad where they might be vulnerable and out of place, and because not all support equipment can be readily stored. Roughly speaking, deploying a single wing of tactical air forces every five days (slightly faster than in Desert Shield) could require 40 percent of the airlift. Deploying Patriot missiles would also be burdensome and, potentially, time-urgent. Other possible candidates for airlift would include (1) specialized equipment for allies,

Table 1.1

Airlift Burden of Diverse Units

Unit	Weight (tons)[a]	C-17 Loads
Airborne ready brigade task force	7,200	80
Aviation brigade task force	6,500	133
Mechanized battalion task force	5,000	68
Mechanized heavy brigade task force	23,000	330
Armored cavalry regiment	30,000	402
MLRS battalion (3 batteries and HQ)	4,000	58
Patriot battalion	4,000	58
MLRS battery (9 vehicles, 114 resupply vehicles, 684 missiles)	1,300	18
HIMARS battery	800	11
Postulated early-entry force (light mobile-infantry component)	5,000	7

[a]Some data inferred from DSB (1996b, p. 37), assuming roughly 70 tons per C-17 load for heavy units and 50 tons for aviation units.

NOTES: MLRS = Multiple Launcher Rocket System. HIMARS = high-mobility-artillery rocket system.

(2) special support for a Marine Corps MEU or MEB (on top of what is onboard amphibious or prepositioning ships), and (3) theater-level command, control, communications, computers, intelligence, surveillance, and reconnaissance (C4ISR) equipment.

Table 1.1 lists some airlift burdens for a variety of units.

The conclusions here are stark:

- It is possible to have brigade-sized light mobile infantry units or mechanized battalions that could be deployed within three to four days by projected airlift (longer if they had to avoid major airfields because of concerns about weapons of mass destruction).

- However, it is extremely doubtful that larger and more mechanized units could be deployed on such a time scale, even with *heroic* efforts to develop lighter tanks, armored personnel carriers, and so on: Moreover, if we could cut weight in half, the airlift times would still be more than a week.

- Given these facts—the high costs of additional airlift, the limited number of airbases that are sometimes available for use, and the increased vulnerability of those bases—it would seem that we should look to other forms of mobility for all but the early allied-support and light mobile-infantry portions of our postulated early-entry force. That is, *we should not expect a light mechanized unit to come in by airlift*, even with substantial force lightening.

A few cautions are worthwhile on terminological ambiguities. Our "light mobile infantry" includes many trucks, some lightly armored wheeled vehicles (in the midterm, current "light armored vehicles" (LAVs), and in the longer term, more-advanced vehicles), and some helicopters. Moreover, in cases in which a large fraction of early airlift could be devoted to deploying ground forces, or in which more warning time were available, our light mobile-infantry component could have many light armored vehicles and helicopters, making it closer to what others describe as an "aero-mechanized force" (Gordon and Wilson, 1998). In the cases we emphasize, however, such capabilities would largely be deployed as part of our light (or medium-weight) mechanized component. Ultimately, the appropriate joint-task-force configuration would depend on details of the situation.

What Is Feasible With More and Faster Sealift? The limitations of airlift are well enough known that many people have suggested that DoD should turn to sealift for the task—using modern ship technology to achieve much faster speeds (e.g., 60 knots or more, rather than 20), or even to employ a hybrid airship, which might operate at 125 knots.[18] With back-of-the-envelope calculations, one might believe that deployment times could be cut to remarkably small numbers (e.g., a few days with an airship, or six days with a 60-knot ship). Both speeds are technologically feasible. This said, we believe that *real-world* deployment times would likely be longer because of loading and unloading, en route refueling (which becomes more necessary

---

[18]Other kinds of hybrid airships have also been built and demonstrated. For example, the Soviet "Caspian Sea Monsters" can fly to altitudes of thousands of feet, but they are intended to cruise upon an air layer just above the ocean.

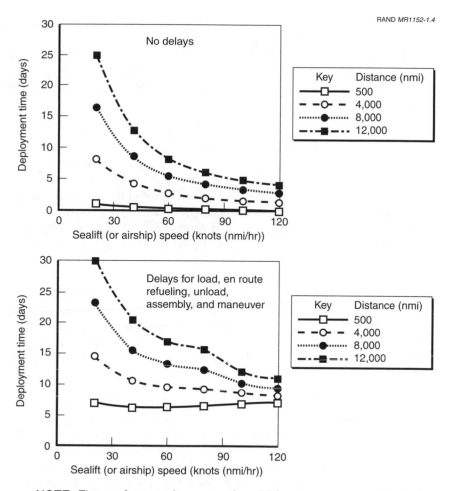

RAND *MR1152-1.4*

NOTE: Figures for speeds greater than 80 knots are assumed to be for airships with one day spent on en route refueling.

Figure 1.4—Parametrics of Deployment Time Using Sealift or Airships

as speeds and distance increase), assembly time, and maneuver from points of debarkation to the combat area. Figure 1.4 makes the point parametrically, assuming 4, 1, 1, and 1 days, respectively, for these delay times. The en route refueling time is for a 30-knot ship going 6,000 nmi; other times are scaled from this in proportion to (speed/30)(distance/6,000). While this is a very simple model, it makes the point that deployment times less than a week will be extremely difficult to achieve—even with postulated high-speed vehicles. A limiting factor will likely be the various delays, which historically have been "stubborn facts" that have been difficult to eliminate.[19]

Our conclusions here are supported empirically. Since 1980, the Marine Corps has gained considerable experience with its Maritime Prepositioning Force (MPF) operations. Because the Marine Corps maintains high readiness for such operations, the planning factor usually quoted for employment is 10 days, limited by where the ships are based (e.g., Diego Garcia or Guam). However, the Marine Corps has demonstrated the ability to maintain ships on station. And with ships already in the region, or by using strategic warning to move to the region early, employment would be possible in a week or slightly less (more than the time to airlift personnel). Another data point here is that the Army today maintains afloat prepositioned equipment in the Persian Gulf. With unusual prior preparations, the Army can also employ a brigade in roughly a week. These data points motivate the bottom curve of Figure 1.4, which assumes that ships are only 500 nmi from their port of debarkation at the time of full-scale deployment.

One further point is important here: In some contingencies (especially larger conflicts), the early-entry forces would not be able to enter administratively through major ports because those ports would be vulnerable to attacks, including missile attacks with chemical and biological weapons. As a result, the unload times would be significantly greater than traditional planning factors. Unloading might need to be "over the shore," a well-known but complex and trouble-prone process. Alternatively, more distant bases could be used with the need for subsequent long-distance maneuver.

---

[19]These limitations are among the reasons that the Army currently has prepositioned equipment on the ground in Kuwait and Qatar. Such equipment, of course, is vulnerable to attacks by weapons of mass destruction.

Conclusions on Mobility

Zero Basing the Way We Use Existing Ship-Based Prepositioning.
Our conclusion, then, is that we cannot depend on rapid sealift to
deploy light (or heavy) mechanized forces within a matter of five to
eight days. Instead, to maximize rapid deployment we are driven
inexorably to the following conclusions:

- The focus of attention for first-week deployment of light mechanized forces should be prepositioning ships. Even one or two
  such ships could carry the bulk of the equipment for the entire
  19,000-ton light mechanized force that we describe in Chapter
  Two.[20] Components of the early allied-support force and pilots
  for the lift helicopters might also be on these ships or Navy
  amphibious ships.

- These prepositioning ships should either be located in the region
  of concern or deployed provisionally upon *strategic* warning.

- If we are to depend on using strategic warning (i.e., in using
  prepositioning ships not in the immediate region, but in places
  such as Diego Garcia or Guam), U.S. military policy should be
  changed so as to develop a pattern of *routinely* deploying such
  prepositioning ships to the crisis region, precisely as the United
  States has deployed aircraft carriers upon such warning.

Less plausibly, similar actions are possible with current fast sealift
(SL-7s) based in the United States. With a week of sailing prior to
D-Day (possible only with enough firm strategic warning), most of a
heavy division and its support could close within about two to three
weeks.

Flies in the Ointment. These suggestions involve more wisely using
existing assets and similar mobility assets that are already being procured or could be procured in the near term. This is in sharp contrast
to postulating massive expenditures on future super-high-speed
sealift. Some such sealift may well be desirable in the long term (and
increased intra-theater lift is almost surely needed for reasons not

---

[20]USMC (1998b, pp. 78, ff) states that a Marine Prepositioning Squadron consists of
four to five ships with equipment for a brigade-level Marine Corps air-ground-task-
force (MAGTF) operation sustainable for 30 days. This also requires significant airlift.

discussed here), but we need not wait for that to occur and it would be imprudent, in our view, to assume that procurements could proceed at a fast pace.

The most important downside to our suggested strategy is that there are substantial costs involved in maintaining prepositioning and fast-sealift ships at a high level of alert and actually sailing them from time to time in response to ambiguous strategic warning. There could be significant issues related to using active military, reserve military, or commercial personnel for operations, maintenance, and the crisis-time deployments.[21] If crises arose often enough, perhaps even half as often as carrier battle groups are dispatched to crisis regions, then the costs to maintain such readiness would be much higher than if one assumed the ships would be employed only in very occasional national crises of a dire nature (e.g., once in a decade).

We have not estimated these additional costs yet, but we have two observations. On the one hand, we expect them to be quite low in comparison with the cost of procuring and operating super-fast sealift or, even worse, doubling the size of our airlift. On the other hand, because of stovepiped programming and budgeting in which it is difficult to move money strategically across categories of activity, we anticipate difficulties. The costs of reworking the ship-based prepositioning efforts, additional operations-and-maintenance costs, and implications for personnel policies might all be noxious to the services and difficult to "make happen" without strong DoD encouragement.

THE NEED FOR A RELATED OPERATIONAL CONCEPT

In this chapter, then, we have stated our case for a first-week intervention force that would include both light mobile-infantry and light mechanized components. Further, we have made the case that such a force could be deployed on a timely basis even if full-scale deployment did not begin until D-Day. The next issue is how to conceive

---

[21]The issue of costs may be less significant than this discussion suggests, at least for Marine Corps prepositioning, because current Marine Corps practices maintain high readiness.

the force itself and, of course, its operational concept. That is the subject of Chapter Two.

OPERATIONAL CONCEPT

JOINT OPERATIONAL CONCEPT

In this chapter, we focus almost exclusively on an operational concept for a stressing case: early intervention in a major theater war that involves an armored invasion of a friendly country. The operational challenge is bringing about an *early halt*, which to us means the first phase of blunting and defeating an armored invasion. Here we envision extremely rapid *joint*-force actions once the decision to act has been made. Figure 2.1 illustrates our ambition for a case in which only minimal and possibly covert activities are possible prior to D-Day. Figure 2.1 is incomplete, but it makes the point that we have in mind *early* and highly parallel joint operations such as connecting with allies, establishing theaterwide defenses, conducting strategic bombing, and reinforcing allied ground forces with units capable of early offensive and defensive operations that could begin immediately—with the benefit of long-range fires in the form of aircraft, missiles, and even naval gunfire with advanced munitions—to unravel the invader's forces and, ideally, bring about his immediate collapse.[1] At a minimum, the intention would be to halt the advance, weaken the opponent's command and control,[2] and set up conditions for follow-on reinforcements and counteroffensive operations.

---

[1]This ambitious objective would be consistent with what some authors call achieving "rapid dominance" (Ullman and Wade, 1996; see especially pp. 14–15).

[2]Some parts of the operation would involve information warfare in various forms. We do not discuss such matters in this study, but see Khalilzad and White (1999) for a broad unclassified discussion. See also Arquilla and Ronfeldt (1997), including a striking historical example (p. 35, from their chapter "Cyberwar Is Coming,") involving the 13th-century Mongols' ability to dominate information and, in so doing, leverage

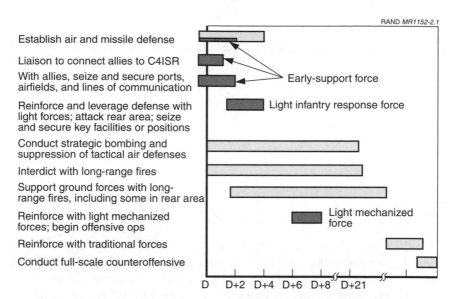

RAND MR1152-2.1

NOTES: The darkest shading is for early-entry ground-force action. Portions would come in earlier from a MEU for the light mechanized force.

Figure 2.1—Operational Concept for Rapid Defense of an Ally

Although the task force operations would involve air, land, sea, and space components, our focus in this monograph is the early ground force. Figure 2.1 indicates key missions for that ground component in the darkest shading. Let us now turn to the ground-component operations in more detail.

Figure 2.2 indicates schematically a composite structure for the joint task force. It emphasizes that we will be focusing primarily on a concept of the ground component, which itself is likely to be joint.

---

their numbers. Although the analogy has been drawn with major theater wars in mind, it may also apply strongly to quick efforts to halt ethnic cleansing operations. For a review of network-centric operations, see Alberts, Garstka, and Stein (1999).

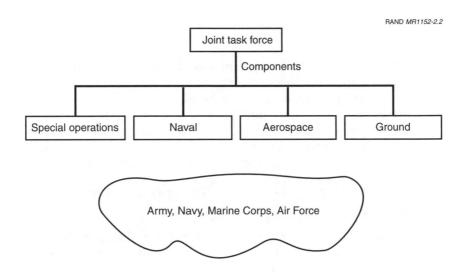

RAND *MR1152-2.2*

NOTE: The JTF commander would resolve command relationships.

Figure 2.2—Joint-Task-Force Composition

THE EARLY-ENTRY GROUND-FORCE CONCEPT

Overview of the Early-Entry Ground-Force Concept

Our concept for a rapidly employable ground force has three components: (1) an early allied-support force; (2) a light mobile-infantry force; and (3) a light mechanized force. Some might add adjectives such as "response" or "strike" to these titles. Some might distinguish sharply between Army and Marine Corps units, but we have consciously rejected doing so for several reasons:

- The mix of Army and Marine Corps units should depend on situational details such as whether Navy and Marine Corps forces had been able to build up in the region during strategic warning.

- Force transformation is and should be to some extent competitive, and we do not want to prejudge which service will prove more successful in developing the various capabilities we describe.

- We think it likely that in many contingencies a mix of Army, Marine Corps, and Special Operations Forces (SOF) units will in fact be optimal for early-entry operations. For example, we suspect that rapid operations from the sea will continue to be a core competence of the Marine Corps and that the Army's ability to operate from the sea will be quite different. Also, we suspect that it will continue to be the case that the Army will have more of the expensive systems such as long-range tactical missiles akin to the Army Tactical Missile System (ATACMS)/Brilliant Anti-Armor Munition (BAT).[3]

- We believe that some difficult rethinking will ultimately be necessary to make sense of which service has what kinds of operational capabilities, and to follow up with necessary changes in doctrine and programming. To speculate now about such matters would be premature.

Our decision on this matter creates some difficulties because it means that we must refer to a larger number of emerging and notional concepts, platforms, and weapons, but we believe it is appropriate nonetheless.

As a whole, then, our conceptual early-entry ground force would be smaller than a current-day light division and would number on the order of 8,500 personnel (Figure 2.3). It could be about 50 percent larger if, for example, it contained one Army and one Marine Corps light mechanized unit. It would be the ground component of a "vanguard force" that would also include Air Force, Navy, and additional SOF units.[4] The entire early-entry ground force would be deployable within about a week, whereas follow-on forces would continue to deploy for weeks, and possibly months.

Let us now describe the components of the early-entry ground force separately.

---

[3]We note, however, that ATACMS, or a longer-range version with reduced payload, could be used from ships.

[4]Others have written about various vanguard-force concepts over the years. See, especially, Blaker (1997) and INSS (1997).

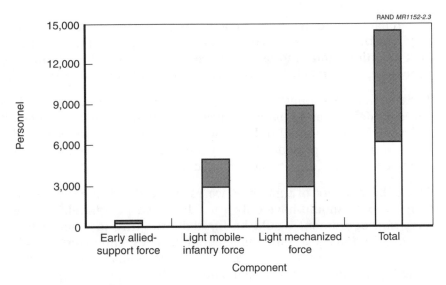

NOTE: Uncertainty is shown in gray.

Figure 2.3—Approximate Composition of the Early Ground Force

The Early Allied-Support Force

The first component to enter—the early allied-support force—would have relatively little in the way of weapons, but would instead (as suggested in Figure 2.1) focus on liaison, C4ISR, and air and missile defense. It should be a rapidly deployable, small-footprint joint force that could be in place with the first round of airlift or by tactical insertion from forward-deployed ships.[5] Once inserted—perhaps during a period of warning before D-Day—this support force would coordinate the U.S. response with allied efforts. It would integrate "first-hour combat power" (U.S. air, naval, and long-range ground fires), provide intelligence, and enhance command and control. As part of this, assuming that the United States has developed a comprehensive information grid allowing its forces to be superbly informed and networked, the early allied-support force could help allies plug in to that information grid to the full extent appropriate.

_____

[5]One option here would be to configure part of a MEU for this function.

This might require the early allied-support force to provide not only expertise but also specialized communications equipment: Despite best intentions and exhortations, those the United States will intervene to help will not always have forces interoperable with its own.

The early allied-support force might also enhance allied ground-based air defense and missile defense. This force would be designed to leverage the potential of landpower (and perhaps airpower and seapower) already present—typically the forces of the U.S. ally being defended. Much of the leverage would depend on C4ISR.

Given this concept of functions, we estimate that the early allied-support force should have only a few hundred personnel.[6] In the long run, equipment would be limited primarily to small arms, communications gear, unmanned aerial vehicles, and vehicles such as the postulated future reconnaissance vehicle (FRV) or even something lighter (and less protected). Patriot-follow-on units might also be necessary, although we treat that deployment as a separate issue that would also be a drain on airlift.[7] In the near to mid term, equipment would be less capable, heavier, or both.

The Light Mobile-Infantry Force

Concept of Operation. Once on the ground, the light mobile-infantry force would be capable of securing a few critical facilities/terrain

---

[6]One way to understand this figure is to imagine that each of three allied divisions has 10 battalions and that the United States has about three advisors in each battalion headquarters, as well as larger numbers of advisors in several higher-level headquarters for air, land, sea, and joint operations. If we add to this an allowance for some special operations teams operating in the field (e.g., behind enemy lines), then the overall number would be in the low hundreds.

[7]The FRV has been described as follows (U.S. Army, 1996): A postulated stealthy ground vehicle that would provide protection of the current M113 at one-third the detectability. It would carry a four-man crew over an extended range of 500 miles. It would be armed with a Javelin or a follow-on to the TOW missile (FOTT) and would have the Objective Crew-Served Weapon (OCSW). Its maximum speed would be 90 kph with 500-mile endurance. Its sensor range would be 4 km. An even lighter version, termed by several agencies the AHMV (Advanced High-Mobility Vehicle) would have many of the same capabilities but be only 2 to 2-1/2 tons in weight and small enough in dimensions to fit internally in a V-22. It would have a "squatting" suspension to allow stackability during strategic or tactical deployment. Its light weight would be achieved at the expense of protection. Passive or active protection against missiles or medium-cannon fire appears to be unachievable at these weight levels.

(e.g., key political centers, ports, or airfields), blocking limited armored advances, denying key regions, defending urban areas, and screening other forces—although not all of this simultaneously. Once deployed, the light mobile-infantry force would have limited but significant tactical mobility in the form of light ground vehicles and rotorcraft. Assuming the presence of worthy but much less capable allied forces, a substantial defense would be possible. This force should be capable of being deployed and in operation effectively anywhere in the world within 48–96 hours.

Such light mobile-infantry forces should be able to exploit an important phenomenon of information age warfare: that firepower potential is growing exponentially while advances in maneuver capability are much more incremental. Light forces, equipped with modern weapons, can defend even against near-peer attackers, unless enveloped or attacked from long distances. As will be discussed in Chapter Three, the technology to improve defensive lethality enormously while reducing lift requirements already exists.[8] This cuts both ways, of course: Adversaries are also likely to have very lethal firepower, even if less sophisticated.

Composition. The precise composition of the light mobile-infantry force would, of course, be highly situation dependent. It might include units for traditional dismounted-infantry missions such as defending an important port, airfield, or capital; perimeter security; crowd control; or psychological operations (PSYOPS). It would be able to link to units with long-range firepower such as future ATACMS-like missiles launched from the high-mobility artillery-rocket system (HIMARS) platform (the follow-on to the Multiple Launcher Rocket System (MLRS)). It would include "precision-strike teams" (PSTs)[9] that would operate at the front or in the enemy's rear area to provide information for long-range fires (particularly crucial

---

[8]Qualitative discussions and quantitative analyses of the large improvements provided by future weapons systems can be found in Steeb et al. (1996a,b) and Matsumura et al. (1997).

[9]These precision-strike teams were a focus of consideration in the 1996 Defense Science Board Summer Study (DSB, 1996a,b). They had been postulated earlier by a number of authors, including colleagues Sam Gardiner, Daniel Fox, Bruce Bennett, and Carl Jones who briefed their conceptual work extensively during 1995 under the title "Understanding the Revolution in Military Affairs." The Marine Corps experimented with related concepts in Hunter Warrior (USMC, 1997a,b).

in mixed terrain, or when military targets might be intermingled with civilians). These precision-strike teams would move using small 2–2$^1/_2$ ton wheeled vehicles that can fit on a variety of lift rotorcraft. These small vehicles could carry four soldiers (e.g., driver, gunner, and two-man dismount team). The unit would also include systems for local reconnaissance and surveillance (e.g., unmanned aerial vehicles (UAVs) such as the Marine Corps' Dragon Drone), or even aerostats.[10] Some of the units would have little more than small arms; others would have their own (i.e., would have "organic") short-range precision fires such as the Objective Crew-Served Weapon (OCSW), follow-on to the TOW (FOTT), and beyond line-of-sight systems such as the extended-range fiber-optic guided missile (EFOG-M) to avoid close battles with numerically superior forces.[11]

Networking. All of the force would be networked: The precision-strike teams would probably have access to other C4ISR and would be linked to long-range fires provided by Air Force, Marine Corps, and Naval aircraft, and by Army and Naval long-range missiles. The teams would also be connected well enough to assure good situational awareness and some level of mutual support. The nature of this networking and related "digitization" is not yet well understood, but such an understanding is emerging as the result of numerous Army and Marine Corps field experiments (see, e.g., USMC, 1997a,b and USA, 1999).

Tactical Mobility. The light mobile-infantry force would be fully air deployable: It would contain only such vehicles and organic equipment as could be deployed by strategic lift to positions close enough to battle so that subsequent movement could be by tactical aircraft (e.g., C-130s or their follow-on, or rotorcraft). The exception would be that up to a regiment's worth of the force might be part of an afloat Marine Expeditionary Unit (MEU).

---

[10]Aerostats are lighter-than-air vehicles (such as blimps) that are able to maintain position (usually by a tether), normally either for radio relay or sensing.

[11]The feasibility of greatly increasing the lethality of light infantry has been demonstrated in extensive simulation-based analysis for the Army and DARPA; e.g., Steeb et al. (1996a,b), DSB (1996b), and Matsumura et al. (1997). It has been further supported by field experiments such as the Marine Corps' Hunter Warrior (USMC, 1997a,b) and recent Army experiments (USA, 1999).

It would be essential for the force to be "agile"—i.e., to have significant tactical mobility as the result of a combination of land vehicles (e.g., trucks, reconnaissance vehicles, and wheeled LAVs) and lift rotorcraft. These would not provide protection against direct-fire attacks by tanks or other large weapons. Instead, survivability would depend primarily on mobility, stealth, standoff, use of terrain, and mutually supportive fires made possible by the networking.

Implications for Size. Based on work by the Army, Marine Corps, Defense Science Board, and ourselves, we estimate that the light mobile-infantry force should be sized at about 3,000–5,000 personnel. Some might be in larger units (e.g., 500 personnel each); some might be in much smaller units 50–80 personnel); and still others would be in a variety of support units. Such a force would have substantial capabilities for early operations—especially if it is part of a much larger allied force or if it is dealing with relatively low-level threats such as poorly organized thugs.[12] This force would, however, be highly vulnerable in many circumstances, as illustrated by the 1993 Rangers and Delta-Force calamity in Mogadishu's urban sprawl[13] or by the Russian Chechnya operation.

An Illustrative Light Mobile-Infantry Force. So far, our discussion has been abstract. To sharpen the imagery, Figure 2.4 shows an illustrative structure that builds on an earlier concept, Task Force Griffin (USA, 1996, and DSB, 1996a,b). Figure 2.4 highlights the point that ground forces would have a number of functions to accomplish simultaneously (including self-protection), which is why even a light force would probably not consist merely of some "spotters" for long-range fires as often assumed by technologists.

The terminology of Figure 2.4 is Army oriented, but we could have constructed a similar structure using Marine Corps concepts. In any

---

[12]Details matter here. The military officers we consulted on this had quite varied opinions about operating against "poorly organized thugs." In some cases such a group might cut and run, while in other cases it could be even more dangerous than organized units because of the element of unpredictability. Much would depend on the local population, friendly defense forces, and prior intelligence on the adversary's support structure, which would probably be the focus of attack.

[13]This is well described by a special "Frontline" show from the Public Broadcasting System (PBS) ["Ambush in Mogadishu," 1998] and in Bowden (1999).

RAND *MR1152-2.4*

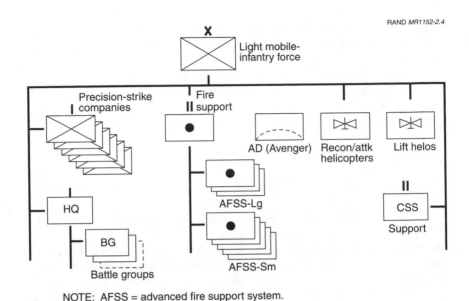

NOTE: AFSS = advanced fire support system.

Figure 2.4—Composition of Illustrative Ground-Component Task Force

case, our illustrative light force is lighter and more quickly deployable than that proposed for Task Force Griffin, because it has significantly lighter vehicles, fewer lift aircraft, and fewer echelons. Reducing the number of echelons (with their associated command centers, vehicles, and staff) from five to three or four should make the force more responsive and flexible for tailoring in different contingencies.

The postulated light mobile-infantry force is composed of five to six precision-strike companies (PSCs), which would call on long-range fires and have their own organic precision capabilities. Each of these PSCs has about five to six precision-strike platoons (PSPs) with 12 Advanced High-Mobility Vehicles (AHMVs), along with several unmanned aerial vehicles (UAVs) for reconnaissance and surveillance. The precision-strike platoons would then be made up of six two-vehicle teams. We estimate that each of the PSCs might control an area 45 km wide by 20 km deep (in open rolling terrain, less in mixed or close terrain). Each PSC has a total of about 300 infantry and almost 100 light vehicles.[14]

The force headquarters includes planning for "battle groups." These battle groups provide flexibility for the force, because they allow this very light force to meet stressing situations by adding more-capable units that may include one or more LAV-based light mechanized companies from the light mechanized force, additional attack helicopters, surveillance systems, air defense, mortars or missile artillery, coalition ground forces, airborne or air assault forces already in place, and other joint units (e.g., a Marine Expeditionary Unit (MEU) that might be present as soon as or sooner than the rest of the light unit.). Battle groups are formed and tailored to deal with specific operational and tactical problems.[15]

The composition of the postulated precision-strike companies is detailed in Figure 2.5. Each is composed of five to six precision-strike platoons made up of six precision-strike teams with two vehicles each. Using two-vehicle teams allows a division of labor and 24-hour operations. Special variations of the basic Advanced High-Mobility Vehicle (AHMV) are designated for command and control (C2), anti-armor, fire support, signals, and UAV ground stations. The direct-fire functions could be accomplished using Javelin, FOTT (follow-on to TOW missile), and the Objective Crew-Served Weapon (OCSW). As mentioned earlier, indirect fire might rely on extended-range fiber-optic guided missiles (EFOG-Ms) and precision guided mortars. Each of these is light and can fit on small vehicles such as the AHMV.

The precision-strike platoons would also have the capability to call in fires from large and small AFSS modules, which could be placed

---

[14]Based on our consultations, we believe that span of control for a precision-strike company, even with the large number of infantry and vehicles, may not be significantly more difficult than for a current infantry company. Digitization, automated planning, and robust communications architectures should allow the commander to direct his team leaders rather than to issue commands to individual soldiers and vehicles by voice or data link. The team leaders themselves should have access to the same planning and execution tools.

[15]Most such "extras" would consume precious airlift, as noted earlier. However, we have thought in terms of using only about 40 percent of that lift for the core "light force," so we are not double-counting.

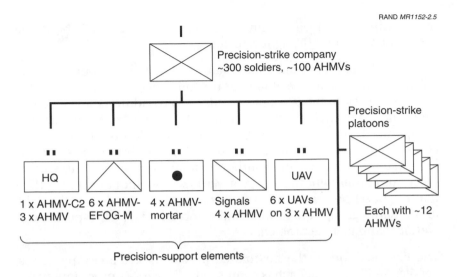

Figure 2.5—Illustrative Composition of a Precision-Strike Platoon

throughout the area of operations by heavy-lift rotorcraft or C-130-type tactical airlifters. The precision-strike platoons would also be capable of calling in long-range fires in the form of naval gunfire, naval missiles, and Air Force, Navy, and Marine Corps aircraft.

We estimate that the light mobile-infantry force might weigh in at something on the order of 5,000 tons—somewhat less than that of the 82nd Airborne's division ready brigade (DRB), but much more than that of a DARPA-postulated small-unit-operations (SUO) force. Table 2.1 shows breakdowns of weights for the DRB and SUO, and a postulated breakdown for our light mobile-infantry force. Our estimate is closer to but smaller than that of Task Force Griffin (DSB, 1996a,b; USA, 1996). It is much larger than the DARPA force because we have sought to allow for more infantry, more tactical mobility for that infantry, and unspecified support, including engineering assets. However, depending on circumstances (including allied capabilities and U.S. experience working with the ally), needs could vary widely. Figure 2.6 makes graphically the same comparisons as in Table 2.1.

Table 2.1

Weight Comparisons for Infantry Units (in tons)

	Division Ready Brigade	DARPA SUO Force	Light Mobile-Infantry Force
Infantry and vehicles	1,633	284	2,600
Command	213	20	100
Artillery/missiles	1,278	1,232	850
Direct-fire systems	710		
Air defense	213		200
Aviation	142		100
C4ISR systems		80	120
Other support	2,840		1,000
Rounded total	7,000	1,600	5,000

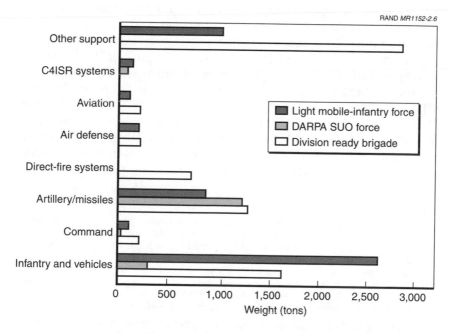

Figure 2.6—Weight Comparisons for Three Infantry Forces

The Light (or Medium-Weight) Mechanized Force[16]

Concept of Operations. The light mechanized force concept (which some would call a medium-weight force)[17] would be a networked, highly mobile force with about five to six agile tactical units with organic lethality and survivability against enemy armor to complement early available standoff fires from land, sea, or air (and the contributions of allied ground forces). Ideally, major portions of the force would be fully air-assault mobile using C-130 or C-130 follow-on aircraft, but some larger equipment (even some tanks) may be needed. The force would deploy primarily by fast sealift or maritime prepositioning (our preferred option), although some of its smaller and lighter units might be airlifted. Light mechanized-force systems would be designed for long endurance and economical expenditure of onboard consumables.

In favorable circumstances (e.g., after long-range fires, with allies aided by the early allied-support force, and the light mobile-infantry force badly weakening and disrupting the attack), the light mechanized force would be able to turn a reactive ground battle against enemy mechanized forces into a proactive seizure of the initiative. Taking advantage of the tactical stability provided by the earlier efforts, the light mechanized force would be able to engage in rapid offensive maneuver. This should be possible even in the early stage of the campaign by exploiting advantages of range, firepower, and information. Although this force's light armored vehicles would be vulnerable to, e.g., high-velocity kinetic energy fires from main battle tanks, its lightweight vehicle-protection technologies (e.g., active-

---

[16]Others have discussed concepts for light, mobile advanced ground forces. Simpkin (1985) wrote of a light mechanized brigade for NATO's use on its flanks or for intervention (p. 290). He emphasized air mobility and the increased firepower available in combat arms (as distinct from artillery). He also recognized that many tactical operations might look more like those of "semi-special" forces rather than traditional units. For more recent discussions, see also MacGregor (1996), USA (1996), and Gordon and Wilson (1998). USMC (1998b) includes sections on military operations in urban terrain and various operational maneuvers from the sea.

[17] The phrase "light mechanized" is ambiguous. Some authors have in mind fewer heavy items of equipment, others have in mind lighter pieces of equipment, and still others envision both. Our description envisions lighter pieces of equipment based on advanced technology, as well as fewer heavy items. A midterm version of the force would, however, be heavier and/or less capable than those envisioned for the long term.

protection systems (APSs)) would defend against lesser hits. The light mechanized force could be inserted in the enemy rear area against supporting infrastructure—provided that it had sufficient joint suppression and support—where it would take advantage of superior situational awareness and long-range, joint precision fires.[18] This force could also fight small armored formations directly from standoff range, and could defend even when outnumbered.

As mentioned above, the light mechanized force would take advantage of the defensive differential of advanced information-age weapons and mobile protected equipment. This force would be much more capable than the light mobile-infantry force in defensive situations in open terrain. The protection provided by the light armor (enhanced in the longer term by new-technology passive and active defenses and by the superiority of situational understanding and the ground mobility of this force's systems) should allow the force to defend in depth with large frontage, even when it is significantly outnumbered. With its operations coordinated with long-range and joint precision fires, it should be able to reduce the enemy's momentum. While it is difficult to know when an enemy would stop its attack, enough lethality can certainly reduce tempo and affect rate of movement. Further, slowing the enemy advance would be critically important to allowing long-range precision fires to operate against combat forces, command-control, and logistics.[19] If the enemy's advance were halted and its forces fragmented, the light mechanized force should—in some circumstances—be able to seize the initiative, regain lost terrain, and set the stage for decisive action by more-potent follow-on forces. This, indeed, is the kind of aggressive and timely maneuver that commanders strive for.

Tactical Mobility. The light mechanized force would gain tactical mobility by using fast, agile vehicles with low-weight protection against indirect fire and small arms, and with tactical engineering assets. Many of the good candidate systems are wheeled, rather than tracked, vehicles. There are trade-offs, of course, because tracked

---

[18]Knowledge of enemy rear area targets and threats would facilitate air strikes on these targets, without risking ground units. Chapter Four shows, however, that unless the targets are in the open and moving predictably, air strikes and other long-range fires can be only moderately effective.

[19]See Davis and Carrillo (1997) and Davis, Bigelow, and McEver (1999).

vehicles have advantages in certain kinds of adverse terrain (e.g., soft sand, mud, snow, and ice). However, advanced wheeled vehicles are much more capable than in earlier years. For discussion see, e.g., Gordon and Wilson (1998) and the references these authors cite.

Lethality. Anti-armor lethality would be in three tiers: (1) long-range precision fires delivered by rockets, missiles, naval gunfire, tactical or long-range air; (2) organic precision indirect and beyond-line-of-sight fires, delivered from weapons platforms in the deployed unit; and (3) rapidly responsive line-of-sight weapons such as hypervelocity missiles and medium-caliber guns mounted on the unit's combat vehicles. Precision guided weapons for these systems are undergoing rapid evolution, as described in Appendix A. By and large, these weapons are within reach in the midterm, based on technology in hand today.

Improved Sustainability. Light mechanized-force systems would be designed for long endurance and economical expenditure of onboard consumables. Whereas today's mechanized brigade in combat might consume 500–1,000 tons per day, this force would be designed to consume something more like 200–300 tons per day, using precision munitions instead of traditional artillery rounds, and employing efficient, high-tech logistics. For example, some supplies could be delivered by precision air drop (assuming control of the air), reducing the necessity of trucks operating over long lines of communications.[20] And, if trucks were used, they also could be much more efficient than today's. Other sources of efficiency are longer-lived batteries, smaller sensors, and low-drain communication systems with smaller radios and relays. Figure 2.7 illustrates estimated resupply rates for the 82nd Airborne and a hypothetical force (again a version of the DARPA small-unit-operations concept. The SUO force requires over 400 tons per day, but much of this is associated with the large advanced fire support system (AFSS)—a 25–ton missile pod). If feasible by virtue of dependable long-range fires provided by aircraft and long-range missiles, the light mechanized force would

---

[20]The Russian Army, during the Manchurian Campaign of World War II, achieved high rates of advance by airlifting fuel for its advancing army. See Despres, Dzirkals, and Whaley (1976).

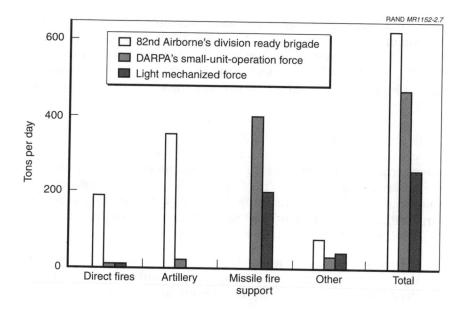

Figure 2.7—Estimated Resupply Requirements for Three Light Forces

require less fire support and should, in any case, be deployable and able to operate effectively within about a week if the relevant sea platforms were in the region or could be deployed with strategic warning.

In some circumstances, the light mechanized force could be used in the enemy rear area and supported there by aircraft operating from ships or relatively distant bases.

Implications for Size. Given the postulated functional requirements and concept of operations, we estimate that the light mechanized force would number about 3,000–5,000 personnel (larger with both Army and Marine Corps versions present, which might well be feasible). Each of the five or six major tactical subunits might number about 500 personnel. Roughly speaking, the mechanized force would have the functionality of at least a current-day mechanized brigade, and the tactical units would have functionality at least comparable to a current-day battalion.

Comment. Such light mechanized forces are not pie in the sky. They are under discussion by both the Marine Corps (often under the rubric of an advanced Marine Corps air-ground task force, or MAGTF, which can be of widely variable size, but can be thought of here as brigade level), and by the Army, particularly in its Army After Next effort and related advanced-warfighting experiments (AWEs). For discussions, see, e.g., USMC (1997a,b), ASB (1998), and USA (1999). It is of interest that recent Army decisions envision reducing the size of even near- to mid-term divisions significantly. These decisions occurred in part as the result of experiments, including some in the desert of the National Test Center (NTC), that helped demonstrate the payoffs of "digitization." This said, a first version of the force within five years would be heavier and less capable than our description. It would be a medium-weight force.

An Illustrative Light Mechanized Force. As with our discussion of the light mobile-infantry component, we present below an illustrative description to provide a departure point for examination.

As shown in Figure 2.8, our longer-term illustrative light mechanized force is an all-arms team integrating the combat, combat support, and combat service support functions down to the lowest possible level.

The future combat vehicle (FCV) is envisioned as a roughly 20–30 ton wheeled or tracked vehicle with both passive and active defenses against incoming rounds and missiles. (A near- or mid-term compromise would be the current 16-ton light armored vehicle [LAV], which currently lacks these advanced defenses.) Passive armor (such as that on LAVs) should defeat medium-caliber cannon fire, and active defenses could defeat enemy smart munitions, explosive anti-tank rounds and missiles, and possibly kinetic energy rounds and missiles. The FCV concept may be modular. That is, the basic chassis would be a common platform with interchangeable systems such as hypervelocity missile pods, multipurpose missile launchers for air defense and indirect fire (as with the current MLRS), an infantry fighting vehicle (IFV) compartment, elevated sensor suite, or

RAND *MR1152-2.8*

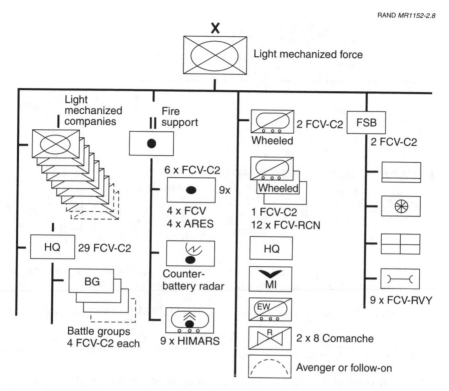

NOTES: FCV = future combat vehicle. RCN = reconnaissance. RVY = recovery.

Figure 2.8—Illustrative Composition of a Light Mechanized Force

engineering components (e.g., for mine clearing, breaching, or recovery.[21]

The light mechanized-force strike company, like that for the light mobile-infantry force, is a relatively self-contained unit. It acts as a combined arms team, integrating all aspects of the force. Additional assets can be attached as needed based on the mission and threat. As shown in Figure 2.9, it has direct and indirect fire systems, infantry

---

[21] Others might include a turreted electromagnetic (EM) gun. Our assessment is that this is still in the distant future for light vehicles because of power-supply issues.

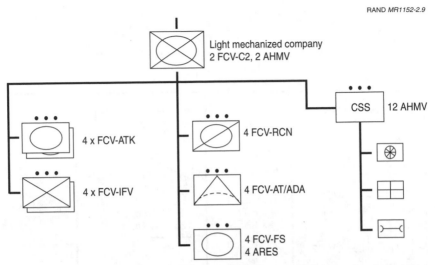

RAND *MR1152-2.9*

NOTES: ATK = attack. RCN = reconnaissance. AT/ADA = anti-tank air defense.
FS = fire support. ARES = advanced robotic engagement system.

Figure 2.9—Illustrative Composition of a Light Mechanized Company

with anti-tank weapons, air defense, engineering, and organic combat service support.

As with the light mobile-infantry force, operations are planned and executed by battle groups formed from command and control cells organic to the light mechanized-force headquarters. Light mechanized companies are attached to a battle group along with additional fire support, reconnaissance, and other available forces tailored to an operation based on consideration of mission, enemy, terrain and weather, time, and troops available (METT-T). Some such companies could be assigned to the light mobile-infantry forces when needed.

Span of control at the company level remains a significant issue. The light mechanized company has nearly triple the weapons platforms of a current heavy (mechanized or armor) company. In addition, the company commander and his subordinates have the challenge of controlling organic reconnaissance, anti-tank, air defense, and indirect-fire assets. However, given a fully digitized force, the commander will be able to send his platoons into battle and orchestrate the fires without time-consuming and unreliable voice communications.

Additional heavy equipment—whether from allies, Marine Corps MEU/MEB, or other Army units—could readily fit into the battle group concept. Likewise, as with the light mobile-infantry force, light mechanized forces would call on long-range fires from all joint and coalition sources to enhance their lethality and shock power.

Reconnaissance-Strike Support Force

Both the light mobile-infantry and light mechanized forces would be supported by at least one light advanced aerial attack/ reconnaissance platform. This aerial platform organic to the joint task force would perform two functions. First, it would provide enhanced scout/reconnaissance capabilities with its own organic sensors. By 2015, these could include third- to fifth-generation forward-looking infrared (FLIR), sensors with some foliage penetration, follow-on to Longbow radar, and possibly electronic and signal intelligence targeting capabilities. Second, it would be capable of engaging targets at a distance with organic firepower such as fire-and-forget advanced Hellfire missiles. In all cases, the aerial platform would be networked with other friendly systems—both on the battle-field and at the theater or national level (e.g., Air Force systems such as the Joint Surveillance Target Attack Radar System (J-STARS) follow-on and satellites). In the near- and mid-term, these functions could be performed by some combination of existing Apache, Cobra, and OH58 series attack/reconnaissance helicopters with capabilities limited to programmed enhancements of today's systems. One obvious candidate for this aerial platform in the future is the Comanche, which can serve both reconnaissance and attack roles.[22] An impor-

---

[22]One of many examples of sensible variations around the structure we provide as a baseline for discussion would be one with substantially more aircraft, particularly the newest versions of Apache and Cobra, and the emerging Comanche. Attack helicopters are highly lethal in the right circumstances. Some officers envision them as somewhat akin to tanks in the sense of being a major source of maneuverable, concentratable firepower. Their vulnerability is also highly situation dependent, however, and the ability to support them well would depend on details such as whether appropriate equipment had been part of the prepositioning sets and whether there were local sources of fuel.

tant consideration here is that Comanches will be self-deployable, thereby greatly reducing the strategic mobility requirements relative to, say, deploying attack helicopters of today's 101st air-mobile/air-assault division. Comanches, however, will still be voracious fuel consumers and have other support requirements.[23] Thus, the concept of operations would be highly constrained by the support requirements and limitations of the major weapons systems involved.

The light mechanized force we have described would weigh on the order of 14,000 tons, as shown in Table 2.2 and Figure 2.10.

A Midterm Version of the Force

A midterm version of the force would be philosophically similar but might depend on the Marine Corps' current light armored vehicles (LAVs) and would likely have at least some heavier tracked vehicles (Bradleys and M-1s). Apaches would be used rather than Comanches. A midterm force would be fully digitized using current and emerging information technologies. As technologies used in weapons, passive and active defense systems, power plants, etc. mature, they could be incorporated into the force to increase lethality, survivability, mobility, and other force attributes. The bottom line is that as technologies mature, advances could be incorporated into the force to enhance capabilities within the force structure concepts we have explored.

SITUATION UNDERSTANDING, INFORMATION DOMINANCE, AND C2

Situation understanding and information dominance are keys to the early-entry concept. As noted elsewhere (DSB, 1996b, pp. 5–6):

> Situational understanding is the key to the rapidly deployable early entry concept. Effective situational understanding permits the independent units to control substantially more territory than their opponents.

---

[23]We limited the number of Comanches in our notional force for this reason.

Table 2.2

Component Weights (in tons) of
the Light Mechanized Force

Component	Weight
Infantry and vehicles	1,600
Command	1,200
Artillery/missiles	1,800
Direct-fire systems	960
Air defense	540
Aviation	200
C41SR systems	1,600
Other support	5,700
Rounded total	14,000

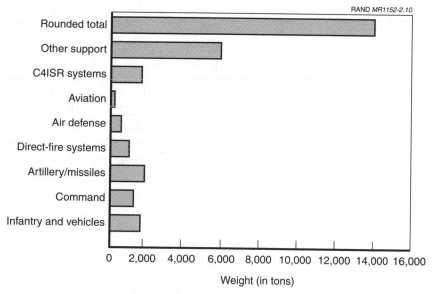

RAND *MR1152-2.10*

Weight (in tons)

Figure 2.10—Weights of Light Mechanized-Force Components

Situational understanding demands automated systems that provide the following information in real-time to the combat units:

- Where am I?
- Where are the other friendlies (this unit, members of other units, supporters, fire support, etc.)?
- What's in the air?
- Where is the enemy (from all sensors), and what is he doing?
- Where are the fixed assets (depots, bridges, etc.) of interest?
- What is my supply and sustainment status?
- What is our plan of operations?

Both the light mobile-infantry and light mechanized forces are expected to contain three layers of sensors, fusion, and automation. Each layer has specific responsibilities, but can also provide backup for adjacent layers. At the top layer, links to theater- and national-level sensors and databases provide sufficient information to plan insertions and prepare for initial combat. These forces also have the ability to gather information about the enemy that is required for planning and execution decisions—such as the placement of subordinate units and the long-range engagement of the enemy. The information gathered should enable the commander to spot weaknesses in air defenses, corridors to insert forces in the enemy rear area when required, places to set up ambushes, and likely targets, and to tell when to extract forces or move to the next engagement or ambush. The midlevel units may also have access to all information available at the next higher level.[24] And, similarly, a reconnaissance organization exists equipped with a variety of sensors to seek information relevant to its operations not available elsewhere. At the bottom level, small combat elements have access to relevant

---

[24]It remains to be seen how much information sharing will prove desirable. Some advocates of network-centric operations consider it almost a matter of principle that the information should be available to all, but we find it difficult to imagine that such a concept will prove desirable in real wars when commanders must be concerned about network security and, in some cases, about keeping information to themselves. Thus, we believe that access will in fact be limited. If networking is to achieve its full potential, however, the normal tendency to compartmentalize a great deal of information should be resisted. Also, the normal dichotomy between a commander's area of influence and his larger area of interest may blur as sensors, communication systems, and weapons all increase range and capability. In this situation, information overload can become a real problem. All of these concerns deserve more extensive study.

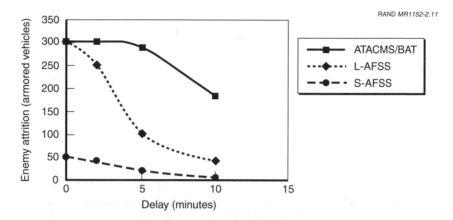

NOTES: These are the results of a high-resolution simulation in a mostly open, mixed-terrain scenario. The systems portrayed in this figure all launch missiles carrying smart munitions. S-AFSS and L-AFSS are the small and large versions of the advanced fire support system, a DARPA concept for small unit operations. S-AFSS is an unmanned pod that launches short-range missiles carrying a single moderate-size footprint submunition, whereas L-AFSS launches medium-range missiles carrying three submunitions each. ATACMS is a long-range missile with many large-footprint submunitions.

Figure 2.11—How Time Delays Can Affect Weapon Effectiveness

information gathered by all the layers above them. They also have a suite of organic capabilities to provide the discrete and time-sensitive information they need to maneuver, and to engage and evade. These organic capabilities include future mini and micro UAVs (some remotely operated and some tethered to manned vehicles) for short-range situational understanding. All will be equipped with advanced multispectral sensors.

Our work on the small unit operations (SUO), Defense Science Board (DSB), and Rapid Force Projection Initiative (RFPI) projects also indicated that *timeliness* is extremely important for weapon effectiveness against fleeting targets. Figure 2.11 shows the effect of even short delays on organic and standoff weapons in a defensive scenario on mostly open, mixed terrain.

All this is predicated on information dominance—the concept that we have a decided advantage in situation awareness—not just that

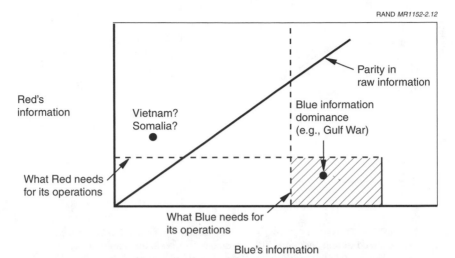

Figure 2.12—Requirements for Information Dominance

NOTE: These are the results of a particular high-resolution simulation for long-range fires operating in mostly open terrain against a mechanized regiment.

we obtain good information. We need (Figure 2.12) to deny the enemy the ability to collect information and—if he does gather it—stop him from capitalizing on it. This will require technologies for jamming, deception, camouflage, and targeting of enemy sensors and C2 nodes, including satellites, headquarters, and land-based communication networks.

CHALLENGES OF COORDINATION IN JOINT OPERATIONS

The challenges involved in employing such an early-entry force would be considerable. The coordination process might be quite complex and might require comprehensive intelligence, precise synchronization, and "network-centric operations." For example, the prepositioning ships would sail to the area of emerging crisis upon early warning. At the same time, military and civilian aircraft would ferry in troops for the light mobile-infantry force. Air operations coordinated with the early support force would ensure air superiority and, if necessary, clear corridors for lift rotorcraft operating from the ships. The lift rotorcraft—carrying equipment such as AHMVs and

AFSS (postulated Army systems) or the Marine Corps equivalent—would link up with the troops at secured airfields and move on to deployment zones. This would all occur in the first days. These light mobile-infantry forces would use their organic fires and might also call in long-range fires from a submarine precision-attack system accompanying the prepositioning ships (see Chapter Three), along with their standoff air-delivered weapons.[25] These fires should block or stall the enemy attack, protect key objectives in the threatened area, and secure lodgments such as airfields and ports. These landing sites would then be used by the light mechanized force, which would then maneuver to exploit the successes of the earlier operations.

The approach would have many advantages despite its complexity. Seeing the strength and flexibility of the sea-based prepositioning forces would deter many aggressive acts and might avoid some contingencies—from police actions to major theater war. The various force components would also have unique self-protection capabilities. For example, the lift rotorcraft on the prepositioning ships (and amphibious ships) could be used for surveillance, minesweeping, anti-submarine warfare (ASW), and other screening operations, while a converted Trident submarine that we discuss in Chapter Three could perform many of the same screening operations against subsurface threats. Some of the missile pods onboard the ships could be loaded with air defense, anti-ship, and anti-missile weapons. Portions of the force should be capable of being inserted by precision "para drop" from strategic lift aircraft. The force would also be capable of air-assault (vertical-envelopment) operations, partially by organic rotorcraft and partially by rotorcraft pooled at higher levels. We note that it will be increasingly important for tactical assault and reconnaissance rotorcraft to be either self-deploying or easily deployed in future large-capacity strategic airlift or fast sealift.

Even if one is quite optimistic about general feasibility—as we are—it seems evident that there is an absolute requirement for extensive research, including purely analytic studies, small-scale empirical

---

[25]There will undoubtedly be major deconfliction problems facing the joint commander in these situations. Airspace deconfliction of aircraft, missiles, UAVs and helicopters may degrade the potential for massing of fires. At the same time, weapon deconfliction is essential to avoiding redundant fires on the same target.

work, and various types of service experiments in a joint context and larger-scale field joint experiments. It should also be evident that all such work (even the field experiments) should heavily exploit modeling and simulation, since it will not otherwise be possible to explore the capabilities in a wide range of circumstances.[26]

RECAPITULATION

The early-entry ground force we postulate would, then, provide a substantial set of capabilities to a joint-task-force commander and do so in the first week after hostilities began. Light mobile-infantry forces would be the first to enter but would be followed quickly by light mechanized forces exploiting sea-based prepositioning (and then more forces would follow). Such a ground force would enable U.S. political leadership to provide effective forward ground presence in politically important scenarios without the risks that faced the 82nd Airborne Division when it deployed in the first days of the 1990 Gulf War.[27] The basic concept for the ground component involves (1) an early allied-support force to connect effectively with allies and provide to their forces some of the benefits of U.S. command, control, reconnaissance and surveillance, (2) a light mobile-infantry force able, within a few days, to secure key points and severely attrit and disrupt (with the aid of long-range fires) advancing enemy units, and (3) a light mechanized force that would initiate maneuver operations early to exploit the results of efforts by the allies, long-range fires, and light mobile infantry. In some cases, it might be possible to rout even a sizable attacking force within a week or so. In other cases the early-entry force would at least buy time sufficient for large-scale follow-up forces.

---

[26]See Davis, Bigelow, and McEver (1999) for discussion.

[27]The 82nd was deployed without organic trucks and could have been either overrun or bypassed had the Iraqis continued south in the first weeks of conflict; it also lacked some basic engineering capabilities such as that for water purification. In contrast, the 82nd participated in the left-hook operation into Iraq during the 100-hour counteroffensive. It did so, however, because trucks and other support had been provided (Gordon and Wilson, 1998, p. 21).

Chapter Three

ENABLING TECHNOLOGIES

Although much can be done within five years using currently available technology, the full development of this unabashedly ambitious force requires continued and increased support to technologies and systems in many areas. Some of these—notably long-range fires, C4ISR, and strategic mobility—are well recognized and have been or are being discussed elsewhere.[1] Here we focus on enablers involving vehicles with advanced protection, armament, and stealthy long-range fires. Many of these technologies are interdependent and rely on common technology developments. All depend on capable personnel.[2]

ADVANCED VEHICLES

Two families of carrying and fighting vehicles may be necessary for the early allied-support and light mobile-infantry forces, and for the light mechanized forces. These units both need a family of vehicles with a high payload-to-weight ratio using lightweight materials,

---

[1]Prospects for long-range fires are discussed in many places, including DSB (1998a,b), NRC (1997c (the weapons volume)), AFSAB (1995 (summary and attack volumes)), the classified 1997 Deep Attack Weapons Mix Study (DAWMS), Ochmanek et al. (1998), and an ongoing RAND interdiction study sponsored by the Joint Staff (J-8) and OSD Strategy and Threat Reduction. C4ISR was the subject of a major classified DoD study in 1997 and a new such study (Information Systems Investment Strategy, or ISIS) has just begun with MITRE, RAND, and Aerospace conducting the study under the sponsorship of OSD (C3I) and the Joint Staff. The Joint Staff (J-8) is also conducting a major strategic-mobility study, which should be out in late 1999.

[2]For speculation about the individual future soldier, see Friedman and Friedman (1996, Ch. 15). This discussion emphasizes what the authors see as the obsolescence of large, massed ground forces while recognizing the continued necessity for infantry.

enhanced energy efficiency, high degrees of reliability, and compactness for strategic lift. Some new trends in vehicle development are hybrid-drive systems (which improve fuel efficiency), active suspensions (which enhance road speed and mobility), composite bodies to reduce weight, and stealthy shapes and coatings to reduce detectability and improve survivability. The lighter equipment—command and control systems, mobile weapons systems, and combat support vehicles and their combat loads—must be transportable by utility helicopters such as the current UH-60 fleet or its follow-on. No loads in the lighter allied-support and light mobile-infantry forces, including logistical transports or their loads, should exceed the lifting ability of the current CH-47 or its follow-on.

Enhanced energy efficiency and high degrees of equipment reliability are important in early operations for two reasons. First, the value of each item inserted into such situations is very high, and failures are difficult to replace. Second, a low consumption rate of parts, fuel, and other supplies means that the inserted force can accomplish more before requiring replenishment. This, in turn, means that a larger fraction of early airlift deliveries can be combat, rather than support, equipment.

The vehicles and their payloads must use space efficiently in strategic lift. Today's units use space inefficiently because their vehicles were not designed as a "family of vehicles" with complementarity and compactness in mind. Their loads tend to exceed space requirements before they meet weight limits. Some vehicles and loads should be stackable and configured to fit the size and dimensions of standard load pallets and containers. Having a common chassis with modular add-ons (protection systems, weapons, stealth packages, C2 units, etc.) allows the burden to be broken up among multiple airlifters. It also allows the force to be configured to the type of mission—e.g., more protection for urban operations, more sensing capability for recon operations, and less payload for maneuver operations.

The longer-term light mechanized force needs a different family of carrying and fighting vehicles—again with energy efficiency, protection, reliability, and compactness, but also with substantial payload for fuel and ammunition, and rapid ground mobility. The purpose of

Table 3.1

Current, Programmed, and Hypothesized Systems

	Current	Programmed	Farther Future
Direct fire	M1-A2 (69 tons)	M1-A3 (70 tons)	FCV (20–30 tons)
APC	M2-A3 (25 tons)	M2-A3 (25 tons)	ACV (8 tons)
Missile launcher	MLRS (26 tons)	MLRS (26 tons)	ARES/AFSS (4 tons)
SPH	Paladin (31 tons)	Crusader (55 tons)	Adv-155 (8+8 tons)
Truck/utility vehicle	HMMWV (4 tons)	HMMWV (4 tons)	AHMV (2.5 tons)
Aviation	AH-64 (8 tons)	RAH-66 (6 tons)	RAH-66 (6 tons)

NOTES: Many of the farther-future concept vehicles are proposed in the Army-After-Next, small-unit-operations, and DARPA programs intended for 20–30 years into the future. The terminology is as follows: FCV is future combat vehicle, a well-protected direct- and indirect-fire combat vehicle; ACV is the advanced combat vehicle, a much lighter and smaller vehicle than the FCV; Adv-155 is a two-part self-propelled Howitzer and reload vehicle set; ARES is an advanced robotic engagement system with a missile pod; AFSS is a stationary advanced fire support system, in the form of a missile pod; and AHMV is the light, fast Advanced High-Mobility Vehicle. Many of these were originally postulated in revolution in military affairs (RMA)–inspired wargames, but many have gone though engineering design reviews. Thus, they are not entirely speculative. APC=Armored personnel carrier. SPH=Self-propelled Howitzer. HMMWV= High-mobility multipurpose wheeled vehicle.

this family of vehicles is to provide a mobile protected ground fighting power tailored to exploit the special conditions encountered early in a counteraggression campaign, or those encountered later in the fluid, nonlinear engagements in the enemy's rear areas. Some of the trends in vehicle development now in progress support this type of effort, as shown in Table 3.1.

Ideally, the vehicles of this force would be transportable by the current C-130 fleet (maximum 22-ton payload, vehicles carried singly or in pairs) and its "super" short takeoff and landing (SSTOL) (30-ton maximum payload) follow-on. No loads in the organization would exceed this limit. This would greatly expand air-insertion opportunities and related flexibility, allowing future "aero-motorized" doctrine and tactics. If the light or medium-weight mechanized force will be inserted from prepositioning ships, however, this requirement could be relaxed. In any case, the equipment would need to be suitable for "over the shore" unloading from ships, since it may or may not be possible to use major ports.

Table 3.2

Trends in Development of Transport Aircraft

Vehicle Payload	Current	Programmed	Farther Future
0–5 tons	UH-60	V-22	JTR
5–10 tons	CH-47	UH-XX	JTR
10–20/30 tons	C-130H	C-130J	SSTOL, tilt-rotor
30+ tons	C-17, C-5	C-17, C-5	C-17, C-5, hybrid airship (aerocraft)

NOTES: Most of the future airlifters shown here are planned, or at least being contemplated, in the Army XXI and Army After Next programs. The UH-XX is a higher-capacity follow-on to the UH-60 helicopter. The JTR (joint tactical rotorcraft) is a proposed 10-ton helicopter. SSTOL is a "super" short takeoff and landing fixed-wing (or tilt-wing) aircraft that would be a follow-on to the C-130. The tilt-rotor is an ambitious version of the V-22, with a much higher payload. The hybrid airship, finally, is a very large (500-ton) blimp/lifting-body concept with relatively low speed operation (125 knots). On the hybrid airship concept, see also an article by Fulghum and Wall, *Aviation Week and Space Technology*, February 22, 1999.

Fuel efficiencies and equipment reliability are as important to this force as to the light mobile-infantry force. Rapid ground mobility and the ability to operate off road are important because protection is derived in large measure from speed, flexibility, and unpredictability of movements. The dispersal enabled by modern digital technologies and networking will itself require the force to be quite agile. Since at least some smaller mechanized units might be airlifted in particular contingencies, the requirement to use strategic lift efficiently would also be important. To the extent possible, cargoes should have dimensions standardized for civil cargo aircraft so that such aircraft could carry the equipment to intermediate staging bases. Equipment should also be designed for rapid transloading from strategic transports to C-130 (or C-130 follow-on SSTOL) aircraft for tactical insertion. Some of the trends for development of aircraft for transporting light and medium-weight vehicles are summarized in Table 3.2.

Survivability of the light vehicles is also of paramount importance. Figure 3.1, based on recent RAND work on light- and heavy-vehicle options, shows what can be expected for different threats. The lower curve is for current U.S. systems; the higher curve shows the potential of active-protection systems, including those capable of

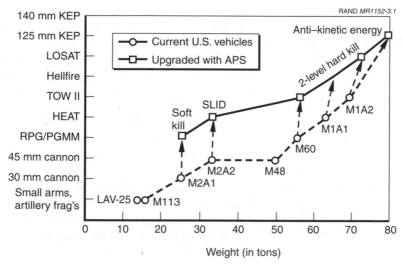

NOTES: KEP = kinetic energy protective system. SLID = small low-cost interceptor device. LAV = light armored vehicle.

Figure 3.1—Effect of Active Protection Systems (APS) for Different Vehicles and Threats

defeating kinetic energy (KE) weapons.[3] By and large, light/medium vehicles of 30 tons or less should in the future be able to defeat moderate-capability anti-tank missiles. Such light/medium vehicles should also provide protection against anti-vehicle mines, either by detection and avoidance, spoofing, or undercarriage armoring. Of course, cost-benefit trades must be made with respect to defending the frontal arc, flank, rear, top, and underside. Additional measures—such as stealth, agility, and special tactics—should further add to vehicle survivability.

_____

[3]The types of improvements derived from advanced protection systems include soft kill (defeat of the missile electronics), SLID (small low-cost interceptor device), 2-level hard kill (collision with the weapon itself and destruction of the warhead), and anti-KE (destruction of a kinetic energy projectile). The types of threats the APS might defeat range from small missiles such as Russian RPGs and precision-guided mortar munitions (PGMMs) to larger high-energy anti-tank (HEAT) missiles comparable to U.S. systems such as TOW II and Hellfire. Very lethal kinetic energy missiles and rounds such as line-of-sight anti-tank (LOSAT) and 125 and 140 mm kinetic energy penetrators fired from current and future enemy tanks require heavy base armor (such as found on 60–70 ton tanks) to survive the KE fragments resulting after the APS engagement. Thus there are limits to what high-tech protection can do for light vehicles.

As noted earlier, both the light mobile-infantry and the light mechanized forces require an advanced aerial attack/reconnaissance platform. It is not clear at this time if this aerial platform should be a conventional rotary wing aircraft (a helicopter) or a tilt-rotor. By 2015, the U.S. military will have extensive operational experience with tilt-rotor aircraft as the V-22 enters the force and gains operational flight hours. The tilt-rotor has several advantages over the helicopter in attack and reconnaissance missions. Its aircraft-like range and speed performance may lead to enhanced self-deployability and cross-battlefield mobility. Its variable rotor-plane orientation capability enables interesting tactical maneuvers in close-combat situations (e.g., backing up in down-slope terrain while still tracking or engaging enemy units). It has disadvantages as well—e.g., larger radar and IR signatures, particularly in the aircraft mode, and uncertain ballistic-damage performance against small-arms fires. An overall system performance and survivability analysis must be made prior to determining which configuration makes the most sense for the aerial platform. In either case, the aircraft avionics suite should include the advanced target acquisition and munitions described above and advanced aircraft survivability equipment (ASE). The ASE should include new radar-warning receivers and jammers, missile-launch detection and tracking sensors, and an active missile-sensor-defeat system such as a laser to blind or dazzle IR missile seekers. In cases of extreme risk, unmanned sensors and vehicles will be required (e.g., unmanned combat aerial vehicles, or UCAVs). These may also take different forms, including tilt-wing and tilt-rotor. Such systems will also need to be inexpensive enough so that losses can be replaced.

ARMAMENT

Both the light mobile-infantry and light mechanized forces should be capable of calling in remote fires from ground, naval, and air systems. Organic to both types of forces would be a follow-on to HIMARS (a high-mobility artillery-rocket system, a smaller version of MLRS) sized for emplacement by CH-47 (and its follow-on) firing large-footprint smart munitions at ranges greater than 60 km. Also present would be several configurations of "rockets in a box," or the Advanced Fire Support System (AFSS). One rockets-in-a-box configuration is sized to be moved and inserted by follower robotic vehicles, light tactical-vehicle towed trailers, or UH-60-class utility heli-copters for 20–35 km range support. The other rockets-in-a-box configuration is sized to be moved and inserted by light mechanized-vehicle towed trailers or CH-47-class heavy lift helicopters for 40–75 km range support. Both systems should be capable of being remotely fired, and both should carry a variety of munitions to include precision engagement of air or ground targets, lethal and nonlethal area suppression weapons, and loitering sensor packages for targeting and target damage assessment. Table 3.3 summarizes some of the trends for these types of weapons over the near and far term. In general, the future systems attain greater range, have larger footprints, and exhibit more target discrimination.[4]

In planning future weapons systems for the ground forces, it is clear that the United States should avoid the classic platform-centric mindset and move instead to network-centric thinking analogous to that reflected, for example, in the Navy's cooperative engagement concept (CEC), which provides a common picture of the naval battlefield to the various ships in a battle group. For ground forces, a network-centric approach would involve networked soldiers, sensors, C2, and weapons with overlapping and complementary capabilities that could be focused adaptively from significant distances (as in "maneuver of fires"). The forces would have both beyond-line-of-sight (BLOS) and line-of-sight (LOS) missiles. Light mechanized fighting vehicles would be equipped with medium-caliber cannons. An advanced suppression weapon, in one of several configurations, would support the requirement for suppression and destruction of dismounted infantry in a variety of environments.

------

[4]Some of the weapons systems listed in Table 3.3 are at the prototype or early devel-opment stage. The electrothermal (ET) and electromagnetic (EM) guns use electric energy (sometimes in combination with chemical energy) to launch kinetic-energy penetrators. The M-Star munition is a large-footprint smart munition, several of which fit in an MLRS-sized rocket. An example is DARPA's Damocles munition. LOCAAS is an orbiting smart munition that can loiter over the battlefield and dive against armored targets it senses. Update-in-flight munitions use a radio or laser link to receive updated instructions on target locations as they fly to the target area. Another example of a loitering system is the postulated unmanned combat aerial vehicle (UCAV), which can drop multiple submunitions on targets it detects. Each of these concepts has distinct advantages and disadvantages for different types of target sets.

Table 3.3
Trends in Weapon Development

	Current	Programmed	Farther Future
Direct fire	120-mm main gun/TOW-2B	Line-of-sight anti-tank missile	Electrothermal/ electromagnetic gun
Organic indirect fire	MLRS with dumb bomblets	MLRS with search and destroy armor (SADARM)/BAT PPI	Advanced Fire Support System/ M-Star munition
Standoff indirect fire	Army Tactical Missile System (ATACMS)	ATACMS with brilliant anti-armor weapon (BAT)	Low-cost anti-armor system (LOCAAS)
Air delivered	Joint standoff weapon (JSOW), Maverick	Wind-corrected munitions dispenser (WCMD)	Update-in-flight/ loiter munitions
Naval fires	Cruise missile, Standard missile	Extended-range guided munition/ naval-TACMS	Update-in-flight/ loiter munitions

NOTE: PPI = preplanned product improvement.

STEALTHY LONG-RANGE FIRES

For the most part, we have omitted discussion of systems for long-range fires because these discussions can be found elsewhere. We do wish to mention, however, the interesting option of using a converted Trident submarine that could accompany a battle group of afloat prepositioned ships. This retrofitted submarine could be ready in as little as six years and would be able to fire hundreds of long-range missiles (a naval version of TACMS, JASSM, Tomahawk, or Standard missiles) from its vertical launch system. Although the Trident option is only one of a number of naval-fire possibilities that include the arsenal ship, missiles launched from many surface combatants, and MLRS launchers on cargo-ship decks,[5] it has the distinct advantage of considerable stealth. Submarine-based strike options may be especially important in the longer run if surface ships become vulnerable, and the midterm Trident option would provide useful experience on which to build. Such submarines could also carry small contingents of ground forces or, e.g., rocket-launched C4ISR systems.

---

[5]See DSB (1998a), pp. 26–29 and Naval Studies Board (1997).

Chapter Four

ANALYSIS

INTRODUCTION

Questions in Search of Analysis

In the preceding chapters, we have sketched a concept of operations and many possible developments involving weapons, systems, forces, and doctrine. Many of the issues raised cry out for analysis because there are fundamental questions to be addressed. Chapter One previewed our analytic findings, but in this chapter we provide more details.

Our analytic work so far has focused primarily on the operational challenge of halting an invading army early. Even for this narrow but very important problem, there are many questions, notably the following:

- How strict are the time requirements that we emphasized at the outset, and on what basis did we estimate them?

- Under what circumstances should it be possible to meet DoD's operational challenge of halting an invasion quickly?

- What can be accomplished, in terms of an early halt, with long-range joint fires alone? How does this depend on detailed circumstances?

- What role should relatively static ground forces play in a halt campaign?

- What role should ground-force maneuver units play in a halt campaign?

- What characteristics should such forces have in order to be sufficiently lethal, survivable, and otherwise effective?

- How do our concepts relate to allied actions?

Background of Past Analysis

It will take years for the defense community to develop a solid understanding of options for future early-entry joint task forces and their ground-force components. Nonetheless, there is a growing body of work generating related insights. In this chapter, we draw upon that work—particularly RAND work with which we are most intimately familiar—to provide some analytical underpinnings for the hypotheses, speculations, and recommendations of this think piece.

The work upon which we draw has addressed the following subjects—sometimes in a broad, exploratory manner and sometimes with highly focused and specialized analysis. As indicated by the citations, different elements of this work benefited from sponsorship by the Army, Air Force, DARPA, Joint Staff, and OSD. The studies we drew upon involved

- strategic mobility for early phases of theater campaigns[1]

- theater-level analysis of major-theater-war (MTW) scenarios under diverse assumptions regarding base access, weapons of mass destruction, warning times, and capabilities of future U.S. forces[2]

---

[1]Davis et al. (forthcoming) (a study for OSD completed in early 1997, which has not yet been granted public release). Also, unpublished work for the Joint Staff (J-8) by Richard Hillestad, Paul K. Davis, Barry Wilson, and Carl Jones.

[2]Davis, Hillestad, and Crawford (1997), work done for OSD, the Joint Staff, and the Air Force.

- halt-phase analysis focused on the potential effects of long-range fires[3]

- alternative concepts for early-entry forces[4]

- defensive potential of light forces enhanced with modern weapons and C4ISR[5]

- feasibility of distributed ground-force operations targeting for and dependent on long-range fires for survival[6]

- potential value of small early-entry maneuvering ground forces as part of a joint task force with long-range fires[7]

- influence of tactical communications on effectiveness of ground-force units with advanced weapons[8]

- influence of enemy maneuver strategy, terrain, weapons characteristics, and C4ISR on the effectiveness of long-range fires[9]

- lethality and survivability of advanced attack helicopters.[10]

- a summary of rapid reaction force studies conducted at RAND over the last six years.[11]

We have also had in mind insights from classified studies by the DoD, MITRE, IDA, and RAND, some in connection with C4ISR, and some in connection with the Deep Attack Weapons Mix Study. In addition, some of us have been deeply involved for several years in work for the Army After Next project. And, finally, we have drawn on

---

[3]Davis and Carrillo (1997); Davis, Bigelow, and McEver (1999); and McEver, Davis, and Bigelow (forthcoming) (for OSD and the Joint Staff) and Ochmanek et al. (1998) (for the Air Force).

[4]Unpublished 1997 work for the Army by colleague Louis Moore.

[5]Steeb et al. (1996a,b) (for the Army and DARPA).

[6]Matsumura et al. (1997) (for the 1996 Defense Science Board (DSB, 1996b)).

[7]Matsumura et al. (1998) (for the 1998 Defense Science Board and Army).

[8]Covington et al. (forthcoming).

[9]Davis, Bigelow, and McEver (forthcoming) (for OSD and the Joint Staff).

[10]Callero et al., unpublished (for the Joint Staff (J-8)).

[11] Matsumura et al. (forthcoming).

insights from the Marine Corps' efforts in the Hunter Warrior and Urban Warrior experiments.

Selected Analysis

With this background, let us now review some of the analytic insights that bear on the subject of this monograph. We discussed mobility issues in Chapter Two. Here we address, sequentially, (1) timeline requirements for an early halt; (2) what could be accomplished by long-range fires alone, or with a combination of long-range fires and ground-force maneuver units; and (3) what implications the analysis has for additional research, including service and joint experiments.

TIMELINE "REQUIREMENTS"

We began this report with an assertion and a notional plot (Figure 1.1) indicating that it was particularly difficult and challenging to halt an invading army *early*, and that capabilities within the first week were critical. We can discuss this more concretely here.

A Simple Single-Scenario Analysis

Figure 4.1 shows illustrative simulation results for one version of a future Southwest Asia (SWA) scenario using an approximation of currently programmed U.S. capabilities and an invasion force with 12 divisions of 700 armored fighting vehicles each, which attack on two axes with two columns each, moving at 60 km/day. There are many other assumptions, notably that the Blue force lacks effective in-place ground forces attempting to hold ground in Kuwait, that suppression of air defenses (SEAD) takes several days, and that tactical air forces may need to operate from long distances and with reduced sortie rates because of concerns about weapons of mass destruction.[12] Although this is only one of many cases (we discuss

---

[12]As of 1999, Iraqi forces are much less capable, the United States has significant forces in the region and others primed for rapid reinforcement using prepositioned equipment, and opportunities for surprise attack are limited. For the purposes of defense planning, however, we may reasonably consider future situations in which the threat to Kuwait and Saudi Arabia is significantly more capable, fewer U.S. forces are

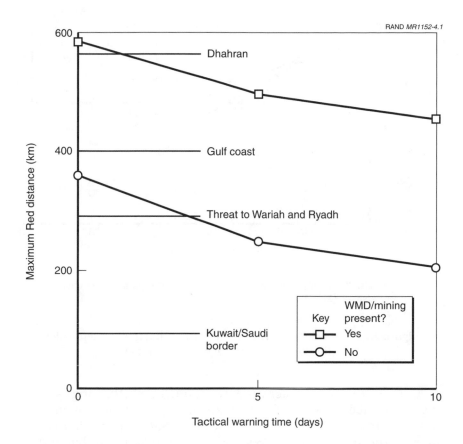

Figure 4.1—Illustrative Simulation Results for SWA Scenario

others later), the results are broadly representative in showing that U.S. capabilities for an early halt (in Kuwait) would be inadequate without significant ground forces in place early. The cases covered by Figure 4.1 show Red penetrations in the range of 200–500 km prior to a halt and counteroffensive (not shown).[13]

---

present, mass-destruction weapons may be used, warning would be ambiguous, access constraints may exist early in crisis, and SEAD is nontrivial.

[13]Related analysis can be found in Davis, Bigelow, and McEver (1999) and Davis and Carrillo (1997). Figures 4.1 and 4.2 were generated by the EXHALT model described in McEver, Davis, and Bigelow (forthcoming). EXHALT is a stochastic, uncertainty-sensitive, desktop model.

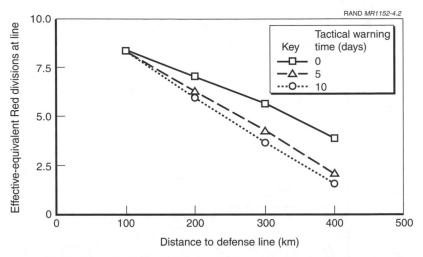

NOTE: Figure includes 12 divisions, with two axes of advance, two columns per axis, 60 km/day, WMD threat, and 1–5 days for suppression of enemy air defenses (SEAD).

Figure 4.2—Red's Effective Remaining Strength (effective-equivalent divisions) at Alternative Defense Lines

Figure 4.2 provides another window on outcomes, this one estimating the number of Red "effective-equivalent divisions" remaining as of the time that Red forces reach what might be possible defense lines at 100 km (in Kuwait), 300 km (northern Saudi Arabia, threatening Ryadh), or 400 km (on the north end of Saudi Arabia's Gulf coast, before the Ras Tanura oil facility). This measure assumes that Red's force loses effectiveness at twice the rate of overall attrition (a common assumption in aggregated models). It is a measure of how large the challenge for ground forces at the defense line might be, again as a function of tactical warning time. Whether we and our allies could have *any* significant force at the defense line, of course, depends on other scenario details such as the extent and use of strategic warning, but Red reaches the defense lines at 100, 300, and 400 km in roughly 2, 6, and 8 days, respectively, more or less independently of tactical warning (not shown). This supports Chapter One's estimates of when we would need to employ the rapidly deployable force postulated in Chapters One through Three. If the early allied-support and light mobile-infantry forces are effective in working with allies to slow the rate of advance, then more time can be bought for deployment and employment of the light mechanized force. Moreover, if long-range fires can be enhanced beyond what is now more or less programmed and assumed in Figure 4.2 (the number of precision weapons actually purchased is a chronic problem), that also would help.

A Scenario Space Analysis

A problem with such depictions is that results depend on so many variables that it is difficult to obtain a sense of the whole. Figure 4.3 considers results from 300 cases chosen, by Monte Carlo techniques, from the set of possible cases created by uncertainty regarding the size of the threat, time required to suppress air defenses, speed of the enemy's advance, his "break point," the spacing of his vehicles, and other factors. The two panels show explicitly the effects of tactical warning time (x axes) and of the United States having to operate from long distances and less well-developed bases because of concerns about weapons of mass destruction (top versus bottom panel). In Figure 4.3 the measure of effectiveness is the "probability" (meaning the fraction of the test set of cases, not a true probability) in which Red is halted at a given distance or less.[14] We see that in no case shown is initial defense in Kuwait (distance of 100 km or less) possible with these currently projected U.S. forces (but see footnote 12), and that the median penetration endangers Ryadh, except for the cases with 10 days' warning time and no WMD threat.

Finally, Figure 4.4 displays comparable information in a simplified "scenario space" somewhat akin to Figure 1.1 of Chapter One. The x axis is tactical warning time, and the y axis is a measure of threat difficulty (here a measure of threat "momentum" relative to a standard force of 10 divisions operating on two axes of advance with two columns per axis each, all moving at a base speed of 60 km/day). Each point in the chart represents one of the simulations from the test set. The simulations are distinguished by their outcome (very bad, bad, marginal, good, or very good) as measured by Red's

---

[14]Subtleties of exploratory analysis, including how to interpret "probabilistic" depictions, are discussed in Davis and Hillestad (in preparation).

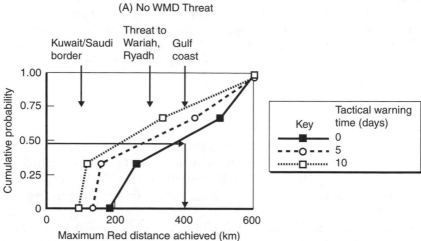

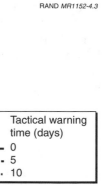

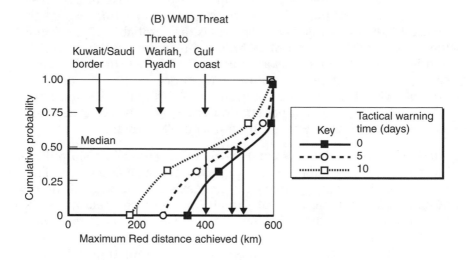

Figure 4.3—Outcomes for a Test Set of Scenarios

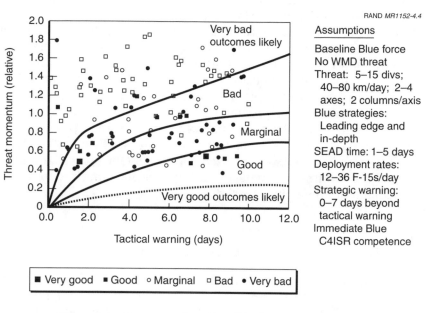

Figure 4.4—Outcomes in a Simplified Scenario Space

maximum penetration distance. The chart shows results only for the no-WMD cases—i.e., an optimistic subset of the whole as indicated in Figure 4.3. The test-set cases use programmed forces and assume significant initial presence, efficient command and control, and some other factors as shown. Although good and bad outcomes can occur anywhere in the chart, very good ones dominate in the lower right and very bad ones dominate in the upper left. We have drawn boundaries to indicate (nonrigorously) typical outcomes by region in the "scenario space."[15]

We see that even with this optimistic test set very good outcomes (roughly, defense in Kuwait) occur only rarely. Good results (roughly, defense in north-central Saudi Arabia) occur only for unrealistically low threat momentum (few divisions, slow movement). Overall, outcomes improve with warning time, but not dramatically for

_____

[15]Outcomes would be more consistent within any given region of the chart if they were for a given duration of the SEAD campaign.

sizable threats. The core problem is that to curtail maximum penetration, the United States would need substantial forces in place, along with quick SEAD and excellent C4ISR. Results are much worse if one allows for weapons of mass destruction, dispersion of enemy forces, mixed terrain with canopy, and other factors. The problem is too multidimensional to discuss in detail here, but suffice it to say, our extensive exploratory analysis of halt-phase capabilities demonstrates that even programmed U.S. forces would have extreme difficulties in future short-warning scenarios with an improved threat (Davis and Carrillo, 1997; Davis, Bigelow, and McEver, 1999). Thus, the assertions of need we expressed in Chapter One are well founded. The United States badly needs to improve its rapidly deployable capabilities for defeating invasion.

WHAT CAN BE DONE WITH LONG-RANGE FIRES, OR A COMBINATION OF THOSE AND MANEUVER FORCES?

Given the need for improved rapid deployment forces, how might the United States proceed? In what follows, we first describe the reasons for not banking entirely on long-range precision fires—despite their enormous value. We then describe analysis comparing outcomes with long-range fires alone with outcomes in cases that combine long-range fires and maneuver forces.

Background on Long-Range Fires

It is quite plausible that in some circumstances long-range fires alone could thwart—and substantially destroy—an invading mechanized army. Many studies and essays have either asserted this or provided analysis that indicates feasibility. The advertised lethality of modern precision weapons is well known.

Unfortunately, the sufficiency of long-range fires is much less clear-cut in many, and perhaps most, cases. The principal reasons are these:

- *Forward Defense Is Often Important.* In most cases of interest, the United States will have strong interest in achieving an *early* halt—e.g., within Kuwait or northern Saudi Arabia, or short of Seoul. Even in the more distant future, it is difficult to identify plausible major theater wars in which it would *not* be important to prevent enemy penetrations of more than, e.g., 30–300 km. Moreover, working with the relevant allies will often make planning for such early halts (forward defense) a political imperative.

- *Fires Alone Have Problems When Depth Is Lacking.* This said, without substantial forces in place, mature command and control systems able to act efficiently from D-Day onward, and an enemy unable to concentrate his maneuvers in time and spread them over multiple axes of advance, the United States would have great difficulty preventing such penetrations (Davis, Bigelow, and McEver, 1999).

- *Rollback, After Penetration, May or May Not Be Sufficient.* In some cases (most notably the Persian Gulf), long-range fires could—after permitting a substantial penetration—effectively "roll back" an enemy (Ochmanek et al., 1998).[16] However, such a rollback might be too late to avoid considerable destruction and might be precluded by political factors—especially once the enemy occupied important cities. Further, such rollbacks would be less feasible against future enemies able to use multiple axes of advance and multiple columns per axis, and to maintain large spacings between vehicles.

- *Fielded Forces Are Routinely Less Capable Than Those in Studies.* To make things worse, when we look at current and programmed U.S. capabilities—as distinct from what is technologically possible and postulated in studies (including ones to which we have enthusiastically contributed)—we find that fielded long-range fires will not be nearly so lethal as often assumed because, for example, the weapons will have relatively poor targeting information. This is particularly a problem when weapons must be launched from long standoff distances because of air defense concerns, and when such weapons cannot be given en route guidance adjusting their impact points and times. Although such en route updates are feasible, they are not generally programmed.

---

[16]As noted by colleague John Gordon, the effect is not literally a "rollback" because, despite high levels of destruction to vehicles, many enemy soldiers could "go to ground" but remain in the area. In other cases, they might flee the area altogether.

- *C4ISR Limitations Can Hurt, Especially With Small-Footprint Weapons.* When the terrain is largely open for targeting purposes, the problem of degraded effectiveness is most severe with small-footprint weapons such as the sensor-fused weapons (SFWs) (e.g., SKEET) planned for the joint standoff weapon (JSOW) and tactical aircraft. Larger-footprint weapons such as ATACMS/BAT do better in this case. Figure 4.5 shows related results graphically and contrasts sensor-fused weapons (assumed four to an F-16 sortie) and large-footprint weapons such as ATACMs missiles fired in two-missile volleys (Davis, Bigelow, and McEver, forthcoming). We see on the top (open terrain with large open areas 10 km across) that for AFV spacings of 100 meters and no delay time, a missile salvo is twice as effective as a four-weapon sortie, whereas with a 20-minute delay the sortie with sensor-fused weapons is an order of magnitude less effective.

- *Canopied Terrain Is Even Worse.* In canopied terrain, even large-footprint weapons lose effectiveness unless their "time since last update" is short (or, equivalently, unless their ability to estimate the enemy's movement speed along the road is very good—despite twists, turns, and possible deliberate changes of speed). Figure 4.5b (bottom panel) shows outcomes for such a case. The effect is particularly sharp when the forces disperse. Table 4.1 shows that the effectiveness of a large-footprint missile can drop by about two orders of magnitude as the result of the enemy's dispersal and increasingly close terrain.

- *Other Countermeasures Are Possible.* A wide variety of other countermeasures can also reduce the effectiveness of fires. These include faster ground-force maneuver (factors of two may be feasible, at least for the first days), long-range mobile air defense systems that delay the use of C4ISR airborne assets and non-stealthy aircraft, intermingling military and civilian vehicles (including hostages), flares, noise generators, and even active-defense systems (DSB, 1996b).

RAND *MR1152-4.5*

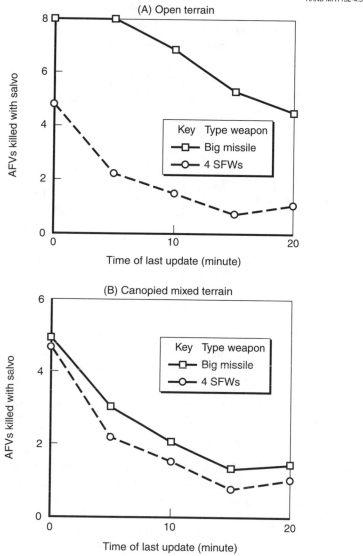

NOTES: The nonmonotonic behaviors are a consequence of our having assumed a particular mean spacing between packets of vehicles. As timing error increases, the probability of hitting the "next packet" increases, even though the probability of hitting the intended packet decreases. Had we considered the inter-packet spacing to be even more uncertain, the nonmonotonicities would disappear.

Figure 4.5—Effects of Weapon Type, Terrain, and Time of Last Update

Table 4.1

Kills per Missile Salvo Versus Dispersal, Terrain Type, and Time of Last Update

0 minutes time of last update			
Dispersal/Terrain	Open	Mixed	Primitive Mixed
Very tight	10	10	4.2
Dispersed	4.2	4.1	1.3
Very dispersed	2.1	2.2	0.62
10 minutes time of last update			
Dispersal/Terrain	Open	Mixed	Primitive Mixed
Very tight	9.4	8.0	2.5
Dispersed	3.3	2.3	0.77
Very dispersed	1.7	1.2	0.38
16 minutes time of last update			
Dispersal/Terrain	Open	Mixed	Primitive Mixed
Very tight	8	5.9	2.5
Dispersed	2.5	1.5	0.19
Very dispersed	1.2	0.9	0.09

NOTES: Details are provided in Davis, Bigelow, and McEver (forthcoming). The mixed and primitive-mixed terrain cases shown assume canopy over most segments of road—as viewed from long slant ranges by an aerial platform such as J-STARS—so that targeting must be directed to occasional "open areas." The figures also assume a 25 percent error in the ability to predict the targeted group's speed along the road over the duration of the weapon's flight (taking into account twists, turns, and changes—perhaps deliberate—in the attacker's movement).

It follows that—although we are bullish about the potential lethality and effectiveness of long-range fires because we believe that the United States could do well in the measure-countermeasure competition over time—we are not sanguine about what capabilities will actually be fielded in the midterm, nor about the certainty of winning the measure-countermeasure game in any *given* conflict.[17]

---

[17]An analogy here might be the 1973 Arab-Israeli war, in which the much superior Israeli Defense Forces were surprised and severely hampered by a combination of good Egyptian tactics and anti-tank weapons on the one hand, and by Syrian surface-to-air missiles on the other. The difficulties the Israelis encountered could have been foreseen and adaptations made (as they were in the subsequent days of the war),

Moreover, we are skeptical about depending on long-range fires alone to halt forces in mixed, canopied terrain, or terrain with substantial urban sprawl.

For these and other reasons indicated in Chapter One, we see considerable value in ground-maneuver forces able to exploit, supplement, and hedge against degradation of long-range fires. Let us now consider concepts for using such forces.

Early Work on Small-Team Concepts

The 1996 Defense Science Board summer study examined a variety of concepts for rapidly deployable light forces. Many of those participating in the study had in mind small teams that would do little more than provide eyes on the ground to assist long-range fires. Others were skeptical about why the eyes on the ground were even necessary, but some of the points mentioned above were at least marginally persuasive. Still others had in mind a broader set of ground-force missions.

Three aspects of this study are relevant here. First, it was evident that long-range fires directed in part by small teams could be quite lethal. Quite a lot could be accomplished by a force roughly the same weight as the 82nd Airborne's division ready brigade. Second, a strong case was made in analysis done for the DSB (Matsumura et al., 1997) that any small "brigade-sized unit" that might be inserted early to stem an invasion should *not* be left to depend entirely on long-range fires for its survival and effectiveness. Even with substantial volume of long-range fires, there would be a significant likelihood of enemy "leakage," which would mean the demise of the small units without an organic close-combat capability.[18] However, advanced but midterm-available organic capability such as the short-range precision fires of EFOG-M and Damocles fired from HIMARS, would

---

but—at the time—the countermeasures to Israeli strength were effective, and Israel had a very close call strategically.

[18]This high-resolution work also demonstrated that large-footprint weapons had major advantages over small-footprint weapons when times from last update are large or enemy dispersal (spacings between armored vehicles) is large, at least in the scenario examined, which involved mostly open terrain.

provide a significant hedge with relatively little weight penalty because no heavy tube artillery was needed for the scenario studied.

Figure 4.6 shows one chart from the analysis. It compares the performance of two forces. The first depends almost entirely on long-range fires for attacking the enemy, while the second assigns the small teams a significant amount of organic short-range indirect precision fire. The forces were designed to produce comparable results and to require comparable airlift. The difference is that, because of leakage, the first force had substantial direct-fire combat in which the small units had to rely on weapons such as TOW and Javelin—a very dangerous situation. In contrast, the second force was able to engage leakers at roughly 20 km—primarily with Damocles fired from HIMARS against artillery and EFOG-M fired against armor. The outcomes for the two forces—in terms of overall loss-exchange ratios—are quite similar. However, note the difference in "shape." Risk analysis (not shown) indicates that because of its ability to engage at greater distances and avoid close combat, the second force's favorable outcome is much more robust. Moreover (not shown), a variant of the second force with no HMMWV-TOWs or AGS, both of which are vulnerable, achieved the same effectiveness while suffering 30 percent fewer losses. Overall, then, the analysis showed high value to the organic precision weapons.

A third important aspect of the summer study was presentation by TRADOC of Task Force Griffin (DSB, 1996b, and USA, 1996), a postulated force that would include small teams to assist use of long-range fires, but that would also include a variety of other units to address other predictable ground-force functions. We (the authors of this monograph) believed that it was a much better concept than the pure small-team approach. As discussed in Chapter Two, Task Force Griffin influenced the ideas we present here.

Analysis Focus for the 1998 DSB Study

Against this background, the 1998 Defense Science Board summer study again addressed early capabilities, but explicitly considered a mix of long-range fires and maneuver forces. In particular, the scenario developed for RAND's supporting analysis with high-resolution simulation was designed with the following in mind:

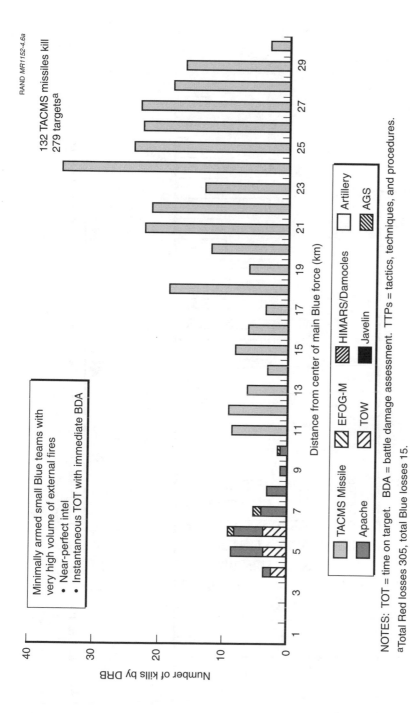

NOTES: TOT = time on target. BDA = battle damage assessment. TTPs = tactics, techniques, and procedures.
aTotal Red losses 305, total Blue losses 15.

Figure 4.6—Comparison of Force Effectiveness: Long-Range Fires with Minimally Armed Small Teams Versus Reduced Long Range Fires with Teams Having Advanced Organic Capabilities

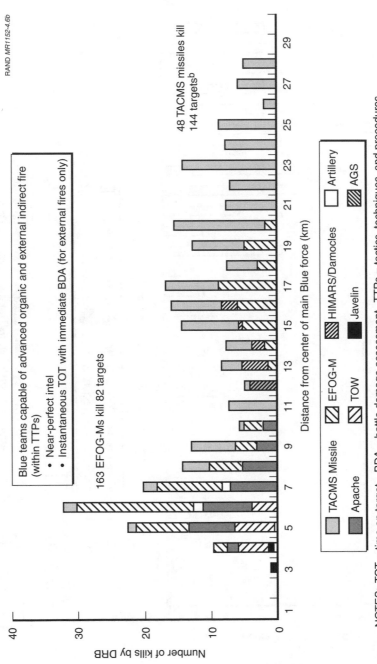

NOTES: TOT = time on target.   BDA = battle damage assessment.   TTPs = tactics, techniques, and procedures.
[b]Total Red losses 297, total Blue losses 14.

Figure 4.6—Continued

- An objective of halting an invasion with light early-entry forces, without the luxury of great depth.

- Mixed, foliated, canopied terrain rather than the more open terrain assumed in most studies (including some, such as DSB (1996) that refer to "mixed terrain," but in fact considered a much more open version of mixed terrain).

- An intelligent adversary who would use relatively easy tactical countermeasures to U.S. precision fires (e.g., dispersion, multiple foliaged roads, sprinting between cover).

- The presence of a defended ally with small but moderately competent ground forces.

- Use of long-range precision fires (LRF) involving strategic and long-legged tactical air forces, and both Army and Navy missiles.

- Interest by the JTF commander in the option of employing small ground-force maneuver units to attack early into the enemy's rear area.

Results of High-Resolution Analysis: Four Joint Concepts Explored

We examined four very different joint operational concepts in a notional 2010–2015 scenario designed to highlight issues associated with phrases such as "information dominance" and "dominant maneuver," and did so for operational circumstances different from those heavily studied in recent years (see Matsumura et al., 1998). The scenario involved early neutralization/disruption of a highly mobile, elite enemy unit with plausible quick-reaction U.S. forces inserted behind enemy lines. That is, the scenario postulated almost immediate "offensive" operations as part of initial U.S. efforts to help the defending ally stop and defeat the invader. All four concepts involved the aggressive use of long-range attack weapons represented by aircraft delivering standoff weapons such as a joint standoff weapon (JSOW) and Navy and Army versions of a tactical missile system (TACMS), which were equipped with advanced submunitions. However, the four joint concepts differed markedly in the level of operational and tactical maneuver with ground forces, and how those forces would be used. All four concepts were examined with a range of assumptions about reconnaissance, surveillance, and target acquisition (RSTA) and about command and control (C2).

Suppression of enemy air defenses (SEAD) was a critical precursor to all concepts, since we assumed that an advanced future threat would respond to U.S. air superiority with integrated air defenses. We did not directly model or simulate this part of the concepts, but rather assumed either (1) that enough SEAD capability would be in place to gain access to the deep enemy battlespace (e.g., successfully clearing an airspace corridor to permit short-range standoff weapons to be delivered and to bring in transport aircraft carrying ground forces, or (2) that the forces were inserted early and then allowed the enemy forces to move past them before beginning their operations.[19]

The first concept explored using long-range, standoff attack alone to neutralize the deep mobile enemy unit. The second concept built on the standoff capability by adding a conventionally organized airborne ground force with updated sensors, C2, and weapons. The third concept used a more agile and more dispersed ground force (sometimes referred to as the enhanced medium-weight strike force) instead of the conventional airborne force.[20] The fourth concept used the same force composition as the third but applied it differently, using the force to attack the relatively "soft" parts of the enemy force rather than the lethal combat units. These last two concepts were reasonably similar to the light mobile-infantry and light mechanized force concepts described in Chapters Two and Three. They were the ones of most interest.

Most of the high-resolution work used a JANUS game board with entity-level representation of armored and other enemy vehicles and a number of specialized models representing the chain of events

---

[19]Since all concepts involve rapid reaction to the invasion in a matter of days, it was deemed unlikely that the entire enemy air defense network could be neutralized. Instead, it was assumed that available SEAD assets would be focused strictly on clearing ingress and egress routes and selected areas of operation. The recent Kosovo operation shows how difficult SEAD can be. Admiral James O. Ellis' briefing on the matter notes that "After 78 days of hard campaigning, we effected little degradation on a modern IADS [integrated air defense system]system."

[20]This concept is one that has many similarities to the USMC's Hunter Warrior, DARPA's Small Unit Operations, and TRADOC's Army After Next Battle Unit and Mobile Strike in that it is a rapidly deployable future force designed around a family of lightweight and agile ground forces.

from sensing to C2 decisionmaking to launcher firing to weapon impact. It represented RSTA and command-control issues parametrically in terms of detection and identification probabilities and delay times.

Case 1: Long-Range Fires Alone. The results of our simulations for Case 1 were quite surprising, even to experienced analysts and military officers (this was before the shortcomings of fires indicated in the last subsection were well understood). We found that long-range, standoff attack alone was largely ineffective in close terrain (mixed terrain with substantial canopy). In retrospect,[21] a belief in the success of long-range fires was due to some common misconceptions: (1) that roads are open areas or, if not because of canopy, that successive open areas will be good detection and killing zones; (2) that detecting any part of a column is enough to project when some part of the column will be at subsequent open areas; (3) that arriving weapons detecting targets in open areas will kill them; (4) that "brilliant weapons" are at least smart; (5) that large-footprint weapons can compensate for moderate errors in predicting target locations; (6) that high levels of fire, even if wasteful, would at least assure kills; and (7) that counter-countermeasures will come along about as fast as countermeasures.

In summary, expectations before the analysis were that effectiveness would be somewhat, but not drastically, lower than obtained for past scenarios with moderately dispersed enemy forces in desert terrain or in mixed terrain without canopy. In fact, as demonstrated by the simulation, vehicles moving on canopied roads with only intermittent open areas (commonly found in "mixed terrain" such as Poland, Germany, and Virginia—except along superhighway corridors) are difficult targets. Predicting future target location is further complicated when, as shown in Figure 4.7, there are multiple minor roads that can be used and the attacker needs to use only some of them. If targets are detected in one open area, then it is unclear where they will be later. A particularly serious targeting problem was the

---

[21]These issues are discussed in detail in Davis, Bigelow, and McEver (forthcoming), which was stimulated by the surprising results of the DSB work, which we did not fully understand at the time despite speculating—with reasonable accuracy—for the report.

Open area where targets
are first seen moving
to right

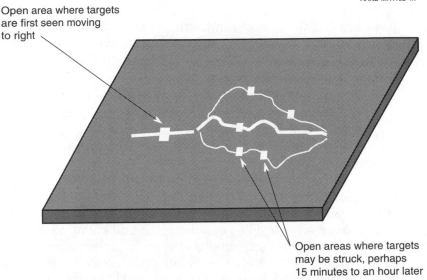

Open areas where targets
may be struck, perhaps
15 minutes to an hour later

Figure 4.7—Intermittent Visibility in Canopied Terrain

enemy's movement pattern with small platoon- and company-sized "packets" widely separated in space and time. This resulted in there being only a few "good" targets in open areas even in those instances in which there were any targets at all (Figure 4.8). If the projected arrival time of packets in the next open area was only modestly in error, a likely result would be no targets at all rather than just another segment of a uniform column. Further, as Figure 4.8 also indicates, an arriving weapon that detected a target in the open might find that its target moved into foliage before it could be struck. That is, the time from detection to impact was in some cases comparable to the transit time across the remainder of the open area. Large-footprint weapons do not solve this problem.

"Brilliant weapons" also have difficulty defeating a dispersed, mobile enemy. Brilliant weapons need to infer the appropriate laydown of their submunitions from multiple vehicle detections. However, if they use an algorithm that lays the submunitions along the centroid, that can in some cases be the worst possible choice if, for example, the centroid line falls in between the two roads along which targets are moving. Fortunately, there are alternative logics. Another serious difficulty at such road crossings is that they often contain a good deal

RAND *MR1152-4.8*

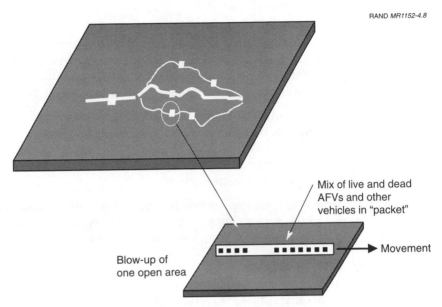

Figure 4.8—Target "Packet" in Open Area

of urban structure, which can confuse the submunitions and substantially reduce kills. Likewise, "barraging" open areas with fire is ineffective because, given the limited number of weapons available, the barrage needs to be well timed. And, equally important, dead targets can build up in the area being barraged and compete with live targets for "smart" weapons, resulting in a rapidly diminishing return for additional shots expended.

Although the analysis was merely illustrative and the outcomes were a sensitive function of many assumptions, several conclusions regarding long-range fires were suggested:

- Expectations for the effectiveness of first-generation long-range fires in mixed terrain against dispersed attacking forces should be low. Nonloitering aircraft could have even more difficult problems if the attacking forces use unpredictable "dash tactics."

- Better information (improved RSTA) and shorter decision times (networking, computer aids, and new doctrine) would be useful but might be much less effective than expected, for long-flight-time systems.

• Short time-of-flight systems (whether fired by maneuver units or, e.g., aircraft) have considerable value. So also, obstruction weapons that can slow or disrupt the enemy movement would be of value, extending the exposure time for use of long-range fires.

• For circumstances such as those studied, "direct-fire" weapons (in the sense of one weapon to one target) are effective not only because of short time of flight, but also because, when operating against a dispersed formation amidst terrain, area-fire weapons are very expensive and inefficient.

• Improved weapon logic is needed for area weapons used against targets in track-crossing patterns, which may be common in the open areas of mixed-terrain regions.[22]

Cases 2–4: Operational Concepts With a Ground Maneuver Component. Cases 2–4 included ground forces, but with different employment concepts. As a result, outcomes were quite different in the simulation. One measure of effectiveness—kills of enemy and losses of the ground force—is shown in Figure 4.9 for Cases 1–4. Each pair of bars in the chart shows, respectively, losses of Red and Blue forces. The last bar-chart pair indicates a representative, postulated, "equivalent" effect from disruption. In the paragraphs below, we describe how each of these results came about.

Case 1, in which air and naval standoff fires were used alone (with no ground forces), caused the least damage to Red for reasons discussed earlier. Targets were difficult to hit with long-range, long time-of-flight standoff weapons.

Case 2, standoff attack with an airborne ground battalion, changed the situation markedly from that of a standoff attack alone. Here the Blue commander used a Marine Expeditionary Unit to establish a lodgment at the coast, enabling an Army airborne ground force (such as an 82nd Airborne battalion) to be inserted against the flank

---

[22]The definition of "area weapon" is ambiguous. A BAT is an area weapon in the sense of killing AFVs in proportion to their density if the number of weapons in a BAT's footprint is not too large. When used against dense formations in the desert, a BAT's effectiveness may be insensitive to the precise target density because of its finite number of submunitions.

RAND *MR1152-4.9*

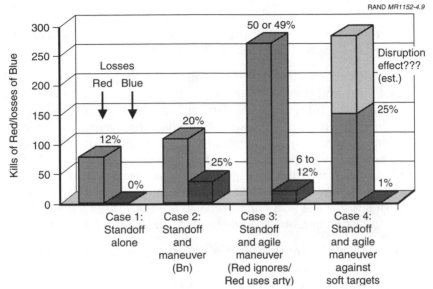

Figure 4.9—Outcomes for Four Cases

of the enemy. The airborne battalion was augmented by two immediate ready companies (IRCs), which each had four M1 tanks and
four M2 Bradleys (deployed with C-17s). By enhancing its apparent
size with deception devices, the IRCs tried to provide a sufficient
threat to turn the lead regiment. If successful, they used a combination of fire and maneuver to try to attrit and disrupt the enemy
attack. In many ways, this was a "conservative" option—i.e., an
extension of current doctrine and tactics.

Assuming that the enemy turns to attack, the results in Figure 4.9
suggest that the ground force could improve somewhat on the
lethality obtainable by standoff fires alone. However, part of the cost
of this additional lethality came in the form of losses to the ground
force. Risk analysis would make Case 2 look even worse.

Case 3 was much more interesting. It represented a stronger departure from current doctrine. Here, there was a deep insertion of
advanced maneuver forces (rapidly deployable 20-ton manned and
robotic ground vehicles) that attacked the enemy forces at many
points, executing ambushes and moving to the next engagement
opportunity. This was done in concert with standoff fires. This

concept was close to that of the light mechanized strike concept we described in Chapters One through Three.

Given a successful insertion, we found that the combination of standoff fires and organic direct and indirect fires was very effective, as shown in the third set of columns in Figure 4.9. The overall lethality of the combination of fires was far greater than for standoff weapons alone. One enemy countermeasure to this operation was to react to the ambushes by placing fire on likely ambush locations with Red artillery. We found that this increased Blue losses but did not significantly change the outcome (because there were many possible ambush sites in the particular scenario studied). Blue's losses went from 6 to 12 percent, but Red's only went from 50 to 49 percent.

Nonetheless, one shortcoming of the agile maneuver force (not reflected in the particular outcomes shown) was its potential vulnerability to massed indirect fires. In Case 4, this was avoided by attacking less dangerous elements such as resupply vehicles, command and control centers, air defense sites, assembly areas, and artillery units. These could have a major effect on the enemy advance (something emphasized by senior military advisors to the study), yet result in few U.S. losses—provided the agile maneuver units could extricate themselves quickly after the attack. Excursions with such a maneuver showed an order of magnitude fewer losses than when attacking similar-sized armor units.

Case 4 therefore emphasized our agile maneuver force as an *exploitation* force designed not to fight the enemy's mechanized force directly, but to neutralize the enemy's close-combat capability. This was achieved by avoiding decisive engagement while destroying his combat-support and combat-service-support infrastructure.

Since the agile ground forces were competing with long-range standoff fires for the same targets—high-payoff logistics and supply vehicles—overall lethality was not as high as seen in Case 3. However, because there was considerably more lethality focused on the support forces, the effect of disruption would be substantially, and perhaps dramatically, higher. How much higher remains to be quantified. Simulation tools, including the ones we used, tend to focus on the more measurable attrition effects. Other effects, such as reduction in morale due to significant losses in short periods of time, tend not to be accounted for. The decision to play Case 4 despite these inherent limitations of current simulations reflected our desire to represent what senior military officers and forward doctrinal thinkers have in mind, even if we had to increase the ratio of subjective assessment to the calculations.

Figure 4.10 summarizes our estimates of both disruption and amount of destruction for the four cases summarized in Figure 4.9. Although semi-schematic, it highlights the fact that disruption and destruction can to some extent be traded off with the same tactical outcome.

Case 1, standoff fires alone, was deemed to have had little disruptive power, because the targets were hit in a more-or-less random fashion. There was little ability to target key elements, hit the heads of columns, or turn the attack. Case 2, with a consolidated force threatening the enemy, managed to turn part of the enemy thrust, delaying it from its original mission. Case 3, involving ambush of enemy combat units, hit specific targets and would probably have created chaos deep in the enemy position. Case 4, involving ambush of enemy soft targets, might well have created even greater disruption by hitting

RAND *MR1152-4.10*

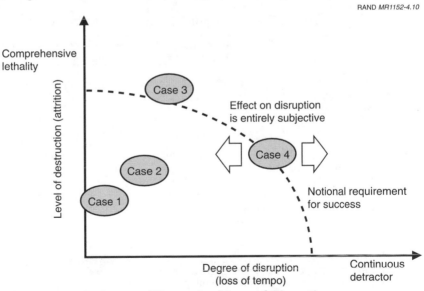

Figure 4.10—Estimates of Enemy Attrition and Disruption

resupply, air defense, artillery, and command and control units in the rear areas.[23]

Time-on-Target and Weapons-Effects Issues. The engagement process employing standoff and organic indirect and direct fires should embody the most efficient combination and sequencing of weapons, taking into account—but not exaggerating—requirements for deconfliction. For example, we found that long-range weapons fired from standoff, which characteristically have large-footprint submunitions, could best be used against large target groups moving predictably and toward open areas. In the close, covered terrain examined for the 1998 summer study, the opportunities for using these weapons were much more limited than expected. Organic indirect-fire weapons could help establish the conditions for other direct-fire weapons and serve also as a means of more robust and selective attrition. These systems could react to smaller exposure intervals than the long-range systems. Direct-fire systems provided quick cycle times and shock, offering the highest degree of robustness and efficiency. (See Table 4.2, which is consistent with the more generic conclusions presented above.)

Table 4.2

Comparisons of Weapon Effectiveness and Efficiency

Weapon class	Time-of-flight (TOF)/ distance traveled	Number of weapons fired/ number of targets killed[a]
Direct fire		
LOSAT	2–3 sec./2–4 km	120/95
Organic indirect fire (w/update)		
Advanced EFOG-M	2–3 min./5–20 km	144/99
Organic indirect fire (no update)		
MLRS-pod (3 subs)	1–2 min./10–40 km	260/65
Long-range standoff fire		
JSOW (2 subs)	10 min./40 km	144/42
TACMS (13 subs)	10 min./150+ km	68/35

[a]Assumptions are those of Case 3.

---

[23]These outcomes might have been different if the enemy had chosen to react more strongly to fires, disperse to an even greater extent, or use decoys extensively. However, these options would have seriously delayed the units closing with the forward lines (their original mission). Furthermore, the assumed Red force was already far more dispersed than any nation's current doctrine and was sprinting between cover. This was a much more "responsive threat" than Iraq's 1991 army.

Other Scenario Variables

Many other countermeasure options might have been postulated for the future threat used in our high-resolution scenario. The invader can disperse its forces, move in unpredictable ways, use deception, employ jammers, launch electromagnetic-pulse (EMP) weapons to prevent U.S. information dominance of the battlefield, use active protection systems, activate counterrecon units, prepare the battle-space, etc. These can reduce the capability of standoff firepower used alone. Maneuver—when feasible—would provide some levels of robustness, allowing some counterconditioning of the battlespace. In addition to threat countermeasures, another obvious scenario variable is weather. It can degrade U.S. overhead and ground sensor capability (especially against stationary targets), deny use of low-altitude airpower, degrade effectiveness of smart munitions, reduce the mobility of maneuver forces, and reduce throughput and timeli-ness of C2, among others. Other less obvious factors include battle-field "friction," fog of war, and systems simply not working as expected. When these happen (and they do, as in Mogadishu), "system" robustness will then be the default judge of force effective-ness. Thus, by our analysis, although standoff firepower has great value, it should be one piece of a larger *joint* operation that includes maneuver forces.

Having emphasized the need for jointness and combined-arms, let us now comment briefly on coalition warfare.

INTERACTION WITH ALLIES

Support of allied capabilities with U.S. C4ISR systems via suitably trained liaison teams should increase the effectiveness of both allied and U.S. ground elements of the coalition force. Three examples strengthen this point:

1. *Operational Level.* The attacker has traditionally been assumed to have a major advantage in mechanized warfare by virtue of his ability to concentrate forces more quickly than the defender can assess the attacker's plan, make his own decisions, and maneuver accordingly. So it is that the famous (or infamous) 3:1 break-even rule for force ratio is said to apply at the tactical level, whereas the

break-even point is perhaps 1.5:1 at the operational level. This merely reflects the traditional assumption that many of the defender's forces will be in the wrong place when the concentrated attack begins. However, with near-real-time reconnaissance and surveillance, it should be possible to greatly reduce this advantage by allowing the defender to assess, decide, and react more quickly. Thus, the value of the early allied-support force and its connections to the U.S. information grid could easily be worth a factor of two in force ratio (Davis, 1995). This is especially plausible if the attacker's air forces and long-range fires are unable to prevent maneuver of friendly forces, and if the defender's in-place forces are strong enough to hold for a day or so awaiting reinforcements.

2. *Missile Attacks With WMD.* It is likely that substantial levels of protection against chemical and biological attack will prove infeasible in practice unless the defender knows when to "suit up"—a nontrivial matter since continual operations in protective suits and masks is not always feasible for long periods of time.[24] Near-real-time warning of missile launches could help in this regard if the related information could be made quickly available to friendly forces on the ground—at least in likely target areas. This could mean the difference between such attacks having catastrophic effects, or merely "serious" effects.

3. *Fires and Special Operations to Slow the Advance.* One of the most critical factors in analysis of the halt phase of campaigns is the speed of the enemy's advance. If friendly ground forces are strong enough to hold temporarily, then advance rates might be low (e.g., 20–40 km/day), but in other cases advance rates might be quite high (e.g., 60–100 km/day)—unless long-range fires in the form of aircraft and missiles, plus special operations on the ground, can create substantial delays.[25] The feasibility of this is greatly enhanced by near-real-time C4ISR and the ability to coordinate

---

[24]One reviewer noted that with new equipment, continuous suited operations are much less difficult than in the past.

[25]In some instances, large numbers of minefields will slow the attacker. These preparations cannot be assumed in most conflicts, however. Furthermore, the rates of advance that we assume in the analysis already have a great deal of slack for operations such as mine clearing.

fires with actions on the ground by small, elite units. Even nations with mediocre armies can have high-performing elite units.

POTENTIAL VALUE OF A NEW RAPIDLY DEPLOYABLE LIGHT FORCE

It is much too early to provide anything like a definitive assessment of what could be accomplished with the kind of rapidly deployable force we advocate in this monograph, but it is useful nonetheless to make some estimates of potential. Figure 4.11 compares results of exploratory analysis with and without the new light force, using almost the same test set of cases presented earlier for the case of a WMD threat.[26] The top panel is the same as Figure 4.3b. The differences between the top and bottom panels are due to assuming that our early allied-support and light mobile-infantry forces could be employed within 3–7 days after tactical warning (longer than mentioned in Chapter One because of assumed delays caused by the WMD threat, which might preclude use of certain bases and ports), that they and allied forces are able to kill about 30–70 AFVs/day, that they and allied forces are able to slow the enemy's advance to the range of 40–60 km/day rather than 40–80, and that the presence of the ground forces effectively increases the quality of the C4ISR used for targeting long-range fires. The test set of cases assumed that the maneuver units could be inserted successfully—by movement on the ground or by rotorcraft insertion, or by reaching positions in advance of the enemy and then allowing the enemy to move through the positions and then be ambushed.

Because so many assumptions are at work, no firm prediction would make sense, but by comparing the exploratory analyses we see that the new light force would have a good chance of being quite significant, even to the extent of making initial defense in northern Saudi Arabia feasible with little tactical warning.

---

[26]Even better prospects for defense in Kuwait are possible for the no-WMD case or if the threat is only minimally capable, as is the case today.

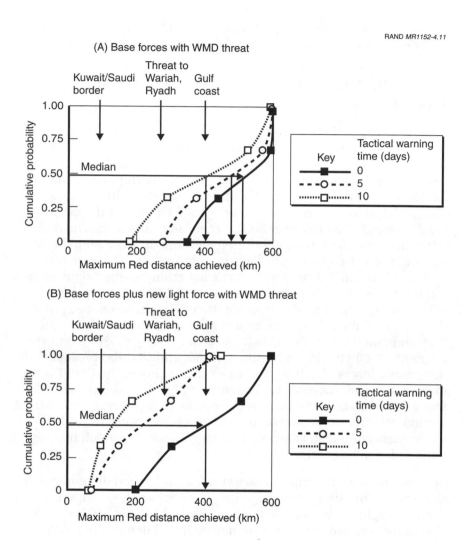

Figure 4.11—Potential Value of a Ground-Force Component
Across a Scenario Space

Table 4.3 summarizes the implications of Figure 4.11 in verbal, quali-
tative terms. For each of several tactical warning times (0, 5, and 10
days), and for the cases with and without the proposed first-week

Table 4.3

Potential Value of a Rapidly Deployable Ground Force for Defense of Kuwait and Saudi Arabia
(outcomes across test set of scenarios)

	Deepest Iraqi Penetration (median from test set of scenarios, followed by a range covering 75% of cases)			
	Without New Ground Component		With New First-Week Ground Component	
Tactical Warning (days)	Median Outcome	Range of Outcomes	Median Outcome	Range of Outcomes
0	Key oil facilities (~500 km)	Gulf coast to Dhahran (~350–580 km)	Gulf coast (~400 km)	Northern Saudi Arabia to key oil facilities (~220–550 km)
5	Gulf coast (~480 km)	Northern Saudi Arabia to Dhahran (~320–580 km)	Northern Saudi Arabia (~200 km)	Kuwait to Gulf coast (~100–380 km)
10	Gulf coast (~400 km)	Northern Saudi Arabia to Dhahran (~200–550 km)	Northern Saudi Arabia (~100 km)	Kuwait to northern Saudi Arabia (~75–350 km)

NOTES: Test set assumes the threat of mass-destruction weapons, which causes reduced sortie rates and deployment rates. The set assumes a threat greater than today's but varies size of threat, base movement rate, number of axes of advance, SEAD time, deployment rates, strategic warning, and tactical warning. For the light mobile-infantry case, it also varies when that force is available, how much attrition it can cause, and how much effect it (and allies) have in reducing the rate of advance.

ground component, the table shows both median outcomes from the test set and the range of outcomes covering the middle 75 percent of cases. As noted above, the postulated ground force would significantly improve odds of a good outcome. It would also decrease the likelihood of a severely bad outcome. All of this, of course, depends on many assumptions.

Significantly, the improved outcomes do not necessarily require success of the small-team concept with insertion deep into the rear area of enemy forces. Analytically, all that was assumed was that the ground forces could cause some attrition on their own and, importantly, reduce the movement rate of the invader's forces. This might be accomplished if ground forces help direct long-range fires such as those from ATACMS/BAT, if allied-force effectiveness is improved with U.S. C4ISR support, or in other ways.

NEEDS FOR EXPERIMENTS AND OTHER RESEARCH

Much of this monograph has been to one degree or another speculative. Some matters can be resolved on the basis of logic, experience, or more formal analysis, but there is a pressing need for a stronger empirical base on which to make judgments about the operational concepts and forces we have discussed. Some of the subjects on which we believe research is needed include the following:

- Concepts of operations for early support forces to work with allies in different missions and different circumstances ranging from humanitarian assistance and counterterrorist actions to major war.

- Ability of ground-maneuver forces (with the assistance of allied forces) to slow rates of advance, even if the invading force has high incentives for speed.

- The ability to assure timely fires—both to assure effectiveness against the invading army and to provide protection for light mobile-infantry forces—taking into account realistic communications, command-control delays, and effects of different terrain and communications gear.

- Logistics to support a first-week ground component, especially the light mobile-infantry component arriving within a few days and, within that when applicable, supporting small teams forward deployed (perhaps behind enemy lines).

- Concepts of operations for successfully inserting small maneuver forces operating at the front or behind enemy lines.

- The ability to achieve a degree of information dominance sufficient to prevent the invader from detecting and attacking the small maneuver units with artillery and longer-range fires.

MAJOR UNCERTAINTIES

No one large critical uncertainty stands in the way of implementing these concepts, but many smaller uncertainties could combine to undermine concept viability, such as the following:

- An inadequate investment in complementary layered sensor suites, automated fusion, reliable communications, and realistic experimentation.

- Acquiring equipment and weapons systems on the basis of the current "platform centric" paradigm as opposed to acquiring equipment and weapons as a "system of systems," which takes into account overlapping reinforcement and complementaries among them.

- Failing to build consensus among important constituencies because the new early-entry force units cannot be judged by previous paradigms and would not always be usable.

- Failing to fund and conduct sufficient testing and evaluation of these concepts and organizations to refine tactical approaches and logistical procedures and planning factors. It will not be enough to make these organizations less resource hungry. Logistical requirements must be accommodated in the initial design of the family of vehicles, including their lift components and their patterns of employment.

- Failing to recognize that the operations of these units will require high performing teams of personnel. Investments in training, training time, and training simulations will be important to leverage the highly capable technologies embedded in these organizations. As investments in technological capability increase, the marginal cost of enhancing human performance has greater payoff.

Although not addressed in the analysis, several problems merit mention. First, mobile defenses available from the Russians and others will provide some capability against currently programmed PGMMs, even TACMS missiles and JSOWs. Second, the ability to accomplish SEAD quickly to permit the operation of aircraft and loitering UAVs may not be easy. Third, while stealthy loitering platforms might be survivable if passive, the act of firing weapons might undercut their survivability. Fourth, decoys and deception may play an increasing role in future conflicts and will affect long-range fires and organic fires differently.

Chapter Five

CONCLUSIONS

Our conclusion is that a joint-task-force approach combining long-range fires with a rapid deployment ground force consisting of an early allied-support force, a light mobile-infantry force, and a light mechanized force has (assuming information superiority and prompt reinforcement) a great deal of potential for major wars and certain kinds of small-scale contingencies. Indeed, the potential is so great that we recommend vigorous efforts, including service and joint experiments, to establish a more reliable empirical base. These should include more stressful field experiments to characterize and assess the leverage achievable by connecting defended allies to U.S. C4ISR systems; the effectiveness and survivability of small teams used at the front or in the enemy's rear area; and the nuts and bolts of inserting, extracting, and supporting such operations with C4ISR, fires, and logistics.

It is possible that a sober empirical assessment will conclude that some aspects of the ground-force concept (primarily the deep insertion of small teams for ambushing) will be too risky for commanders to embrace. But the opposite conclusion is also possible—at least for instances in which the United States is able to prevent enemy surveillance and has a good intelligence on the battlefield. Moreover, other aspects of the concept (e.g., the value of inserting unmanned "missiles-in-a-box" systems and the value of early deploying long-range missile batteries, attack helicopters, and mobile infantry) may prove more robust. In any case, vigorous research, experimentation, and analysis are badly needed.

Although major questions still exist regarding some aspects of the concept, we believe that other aspects are well enough understood that the United States should seek initial versions of an operational first-week ground-force capability within five years. This will require changes in the use of strategic mobility, doctrine, and program priorities. A top priority should be zero basing the use of current maritime prepositioning assets (and airlift) to enable a near- to mid-term version of the advanced joint task force. Initial forces may be heavier and less capable than technology will make possible in the longer term, but much can be done within five years. We believe that establishing such a near- to mid-term goal would be liberating to military innovators, who have arguably been hampered by an excessive emphasis on the distant technology of super-light forces and advanced lift.

BACKGROUND DATA ON SYSTEMS AND FORCE COMPONENTS

Table A.1

Background Data on Systems

Name	Function	Characteristics	Status
Apache Longbow and Longbow/ Hellfire AH-64	Improved attack helicopter (fire and forget; mm wave radar; automatic detection and classification)	140 kt speed; carries 16 Hellfire or mix of missiles and rockets; D version has LONGBOW radar	758 AH-64s to be converted by 2008
Comanche (RAH-66)	Helicopter for armed reconnaissance	170 kt speed; stealthy; can carry 6 Hellfire or Stingers internally	Procurement begins in 2004, totaling 1,292 by 2026.
V-22 Osprey	Marine tilt-rotor aircraft, medium-lift capability	350 kt max speed; can carry 22 troops or up to 10 tons	Marine Corps procuring 425 between 1998 and 2021
MLRS (multiple launcher rocket system)	Heavy, tracked fire support system	Can ripple fire two pods of six MLRS rockets each, or two ATACMS missiles	In current inventory
HIMARS (high-mobility artillery-rocket system)	Medium weight, wheeled fire support system	Based on 5-ton truck; carries single pod or one ATACMS	In prototype testing

Table A.1—Continued

Name	Function	Characteristics	Status
ATACMS/BAT (Army tactical missile system, brilliant anti-armor munition)	Semi-ballistic missile with range in excess of 100 km; BAT effective against stationary and mobile targets	Carries 13 large-footprint BAT submunitions; BAT has both acoustic and thermal sensors to detect targets	Block II ATACMs have BAT and 145 km range; IIA has BAT PPI and 360 km range. 1,200 and 600 missiles will be procured, starting in 2001
SADARM (sense and destroy armor) anti-tank munition	Top-attack anti-tank submunition	Can be carried in MLRS missile or 155 artillery round; small footprint; IR and millimeter wave sensors	In prototype testing
Javelin anti-tank missile system	Replacement for Dragon shoulder-fired anti-tank system	2 km range, fire-and-forget operation; crew of two	In low-rate initial production
EFOG-M (extended-range fiber-optic guided missile)	Short to medium range (15 km) anti-armor missile	Thermal sensor in nose of missile sends image back to operator	In prototype testing
LOCAAS (low-cost anti-armor system)	Loitering platform with anti-armor capability	30-minute flying time; laser radar seeker	In prototype testing
HMMWV (high-mobility multipurpose wheeled vehicle)	Multifunction light vehicle (infantry, recon, anti-armor, air defense, etc.)	4–5 ton weight,	In current inventory
AHMV (Advanced High-Mobility Vehicle)	Multifunction light vehicle (infantry, recon, anti-armor, air defense, etc.)	2–2.5 ton weight	Conceptual

Table A.1—Continued

Name	Function	Characteristics	Status
FSCV (Future Scout and Cavalry Vehicle)	Family of medium weight vehicles	20-ton vehicle carries gun, missiles, sensor package	First unit equipped scheduled for 2008
Predator UAV (Unmanned Aerial Vehicle)	Theater and tactical surveillance, communications relay	25,000 ft. max altitude, 24–30 hr. endurance; carries IR, TV, and radar	In current inventory
Outrider UAV	Tactical surveillance and comm relay	10,000 ft. altitude; 4-hr. endurance, imaging sensor payload	In operational testing

NOTE: Transport aircraft (such as UH-60) are not included in the system summary.

Table A.2

Characteristics of Early-Entry Ground-Force Components

	Early Allied Support	Light Mobile Infantry	Light Mechanized	Total
Functions	Liaison; connections to U.S. C4ISR	Secure ports, installations, cities. Reinforce allied units in key sectors	Block, screen, defend, and counterattack using maneuver and fires	
Personnel	~400	~4,000	~4,000	~8,400
Weight (tons)	100–150	~5,400	~13,600	~19,000
Vehicles	30–50	~900	~800	~1,800
Aircraft	—	8	16	24

BACKGROUND DATA ON UNIT LIFT REQUIREMENTS

Table B.1

Background Data on Unit Lift Requirements

Unit	Manning	Primary Platforms	Tonnage	Daily Fuel and Ammo Tonnage
Air wing	5,000	72	7,000	1,300
Air Expeditionary Wing (est)	2,500	72	4,000	1,300
Patriot Battalion	651	81	15,000	100
AOE MLRS Battalion	132	27	2,400	260
AOE artillery Battalion	663	24	3,300	220
MAGTF (est)	2,711	150	15,000	600
AOE brigade	5,000	400	25,000	800
AAN battle force (est)	6,000	1,400	13,000	300

SOURCE: Estimates given in Army Science Board Briefout, *Concepts and Technology for the Army Beyond 2010*, July 1998.

NOTE: The C-5 and C-17 fleet has a 17,000-ton one-time lift capacity.

BIBLIOGRAPHY

Air Force Scientific Advisory Board (AFSAB) (1995), *New World Vistas: Air and Space Power for the 21st Century*, United States Air Force, Washington, D.C. (multiple volumes; see especially the *Summary* and *Attack* volumes).

Alberts, David S., John J. Garstka, and Frederick P. Stein (1999), *Network Centric Warfare: Developing and Leveraging Information Superiority*, CCRP Publication Series, Department of Defense, Washington, D.C. See www.dodccrp.org/.

Army Science Board (ASB) (1998), *Concepts and Technologies for the Army Beyond 2010*, as briefed to the 1998 Defense Science Board Summer Study in July 1998, draft.

Arquilla, John, and David Ronfeldt (eds.) (1997), *In Athena's Camp: Preparing for Conflict in the Information Age*, RAND, Santa Monica, CA. For RAND publicly available publications, see web site at www.rand.org/.

Barnett, Jeffery R. (1996), *Future War: An Assessment of Aerospace Campaigns in 2010*, Air University Press, Maxwell AFB, AL.

Bennett, Bruce W. (1995), *Two Alternative Views of War in Korea: The North and South Korean Revolutions in Military Affairs*, MR-613-NA, RAND, Santa Monica, CA (limited to government officials).

Bennett, Bruce W., Christopher P. Twomey, and Gregory F. Treverton (1999), *What Are Asymmetric Strategies?* DB-246-OSD, RAND, Santa Monica, CA.

Betts, Richard (1982), *Surprise Attack: Lessons for Defense Planning*, Brookings Institution, Washington, D.C.

Bingham, Price (1997), *The Battle of Al Khafji and the Future of Surveillance Strike*, Aerospace Education Foundation, Arlington, VA.

Blaker, James (1997), *A Vanguard Force: Accelerating the American Revolution in Military Affairs*, Progressive Policy Institute, Washington, D.C.

Bowden, Mark (1999), *Black Hawk Down: a Story of Modern War*, Atlantic Monthly Press, New York.

Bowie, Christopher, Fred Frostic, Kevin Lewis, John Lund, David Ochmanek, and Phil Propper (1993), *The New Calculus: Analyzing Airpower's Changing Role in Joint Theater Campaigns*, RAND, Santa Monica, CA.

Cimbala, Stephen J. (1994), *Military Persuasion: Deterrence and Provocation in Crisis and War*, Pennsylvania State University Press, University Park, PA.

Cohen, Secretary of Defense William S. (1999), *Annual Report to the President and Congress*, Washington, D.C.

Congressional Budget Office (1997), *Moving U.S. Forces: Options for Strategic Mobility*, Washington, D.C.

Cousins, Norman (1987), *The Pathology of Power*, Norton, New York.

Covington, T., R. Steeb, T. Herbert, S. Eisenhard, and D. Norton, (forthcoming), "Exploration of the DARP Small Unit Operations Concept Using High-Resolution Simulation," RAND, Santa Monica, CA.

Davis, Paul K. (1988), *Toward a Conceptual Framework for Operational Arms Control in Europe's Central Region*, R-3704-USDP, RAND, Santa Monica, CA.

Davis, Paul K. (1994), "Improving Deterrence in the Post–Cold War Era: Some Theory and Implications for Defense Planning," in Paul K. Davis (ed.), *New Challenges for Defense Planning: Rethinking How Much Is Enough*, RAND, Santa Monica, CA.

Davis, Paul K. (1995), *Aggregation, Disaggregation, and the 3:1 Rule in Ground Combat*, MR-638-AF/A/OSD, RAND, Santa Monica, CA.

Davis, Paul K., James Bigelow, and Jimmie McEver (1999), *Analytical Methods for Studies and Experiments on "Transforming the Force,"* DB-278-OSD, RAND, Santa Monica, CA.

Davis, Paul K., James Bigelow, and Jimmie McEver (forthcoming), *Effects of Terrain, Maneuver Tactics, and C4ISR on Long-Range Precision Fires: A Stochastic Multiresolution Model (PEM) Calibrated to High-Resolution Simulation*, RAND, Santa Monica, CA.

Davis, Paul K., and Manuel J. Carrillo (1997), *Exploratory Analysis of "The Halt Problem": A Briefing on Methods and Initial Insights*, DB-232-OSD, RAND, Santa Monica, CA.

Davis, Paul K., David Gompert, Richard Hillestad, and Stuart Johnson (1998), *Transforming the Force: Suggestions for DoD Strategy*, RAND, Santa Monica, CA.

Davis, Paul K., David Gompert, and Richard Kugler (1996), *Adaptiveness in National Defense: The Basis of a New Framework*, IP-155, RAND, Santa Monica, CA.

Davis, Paul K., and Richard Hillestad (in preparation), "Exploratory Analysis for Strategy Problems of Massive Uncertainty," RAND, Santa Monica, CA.

Davis, Paul K., Richard Hillestad, and Natalie Crawford (1997), "Capabilities for Major Regional Conflicts," in Zalmay Khalilzad and David Ochmanek (eds.), *Strategic Appraisal 1997: Strategy and Defense Planning for the 21st Century*, RAND, Santa Monica, CA.

Davis, Paul K., William Schwabe, Bruce Narduli, and Richard Norton (forthcoming), "Mitigating Effects of Access Constraints in Persian Gulf Contingencies," RAND, Santa Monica, CA.

Defense Science Board (DSB) (1996a), *Summer Study Task Force on Tactics and Technology for 21st Century Military Superiority, Volume 1, Summary*, Office of the Under Secretary of Defense for Acquisition and Technology, Department of Defense, Washington, D.C.

Defense Science Board (DSB) (1996b), *Summer Study Task Force on Tactics and Technology for 21st Century Military Superiority, Volume 2, Part 1, Supporting Materials*, Office of the Under Secretary of Defense for Acquisition and Technology, Department of Defense, Washington, D.C.

Defense Science Board (DSB) (1998a), *Joint Operations Superiority in the 21st Century: Integrating Capabilities Underwriting Joint Vision 2010 and Beyond, Volume 1, Final Report*, Office of the Under Secretary of Defense for Acquisition and Technology, Department of Defense, Washington, D.C.

Defense Science Board (DSB) (1998b), *Joint Operations Superiority in the 21st Century: Integrating Capabilities Underwriting Joint Vision 2010 and Beyond, Volume 2, Supporting Analyses*, Office of the Under Secretary of Defense for Acquisition and Technology, Department of Defense, Washington, D.C.

Department of Defense (DoD) (1996), *Joint Vision 2010*, Joint Staff, Washington, D.C.

Despres, John H., Lilita Dzirkals, and Barton Whaley (1976), *Timely Lessons of History: The Manchurian Model for Soviet Strategy*, R-1825-NA, RAND, Santa Monica, CA.

Friedman, George and Meredith (1996), *The Future of War*, Random House, New York.

Gordon, John, and Peter A. Wilson (1998*)*, *The Case for Army XXI "Medium Weight" Sero-Motorized Divisions: A Pathway to the Army of 2020*, Strategic Studies Institute of U.S. Army War College, Carlisle Barracks, PA.

Hundley, Richard (1999), *Past Revolutions, Future Transformations: What Can the History of Revolutions in Military Affairs Tell Us About Transforming the U.S. Military?* RAND, Santa Monica, CA.

Institute for Defense Analyses (1997), *Deep Attack Weapons Mix Study*, Alexandria, VA (limited distribution).

Institute for National Strategic Studies (INSS) (1997), "Force Structure," Chapter 21 in *Strategic Assessment: Flashpoints and Force Structure*, National Defense University, Washington, D.C.

Johnson, Stuart E., and Martin C. Libicki (eds.) (1995), *Dominant Battlespace Knowledge: The Winning Edge*, National Defense University Press, Washington, D.C.

Joint Chiefs of Staff (1997), *Concept for Future Joint Operations*, Fort Monroe, VA.

Khalilzad, Zalmay M., and John White (eds.) (1999), *Strategic Appraisal: The Changing Role of Information in Warfare*, RAND, Santa Monica, CA.

Knorr, Klaus, and Patrick Morgan (1983), *Strategic Military Superiority: Incentives and Opportunities*, National Strategy Information Center, New York.

MacGregor, Douglas A. (1996), *Breaking the Phalanx: A New Design for Landpower in the 21st Century*, Praeger, Westport, CT.

Matsumura, John, Randall Steeb, John Gordon IV, Thomas Herbert, Russell Glenn, and Paul Steinberg (forthcoming), "Lightning over Water: Sharpening America's Light Forces for Rapid Reaction Missions," RAND, Santa Monica, CA.

Matsumura, John, Randall Steeb, Thomas Herbert, Mark Lees, Scot Eisenhard, and Angela Stich (1997), *Analytic Support to the Defense Science Board: Tactics and Technology for 21st Century Military Superiority*, DB-198-A, RAND, Santa Monica, CA.

Matsumura, John, Randall Steeb, Ernst Isensee, Thomas Herbert, Scot Eisenhard, and John Gordon IV (1998), *Joint Operations Superiority in the 21st Century: Analytic Support to the 1998 Defense Science Board*, DB-260-A/OSD, RAND, Santa Monica, CA.

McEver, Jimmie, Paul K. Davis, and James Bigelow (forthcoming), "EXHALT: An Interdiction Model for Exploring Halt Capabilities in a Large Scenario Space," RAND, Santa Monica, CA.

National Defense Panel (NDP) (1997), *Transforming Defense: National Security in the 21st Century*.

National Research Council (NRC) (1997a), *Modeling and Simulation (1997), Vol. 9 of Technology for the United States Navy and Marine Corps 2000–2035,* Naval Studies Board, National Academy Press, Washington, D.C.

National Research Council (NRC) (1997b), *Post–Cold War Conflict Deterrence,* National Academy Press, Washington, D.C.

National Research Council (NRC) (1997c), *Weapons (1997), Vol. 5 of Technology for the United States Navy and Marine Corps 2000–2035,* Naval Studies Board, National Academy Press, Washington, D.C.

National Research Council (NRC) (1999), *Naval Expeditionary Logistics: Enabling Operational Maneuver from the Sea,* Naval Studies Board, National Academy Press, Washington, D.C.

Naval Studies Board (1997), *Technology for the United States Navy and Marine Corps, 2000–2035: Becoming a 21st-Century Force* (9 volumes), National Research Council, Washington, D.C.

Ochmanek, David, Edward Harshberger, David Thaler, and Glenn Kent (1998), *To Find and Not To Yield: How Advances in Information and Firepower Can Transform Theater Warfare,* RAND, Santa Monica, CA.

Pape, Robert A. (1996*), Bombing to Win: Air Power and Coercion in War,* Cornell University Press, Ithaca, NY.

Peters, Ralph (1999), *Fighting for the Future: Will America Triumph?* Stackpole Books, Mechanicsburg, PA.

Scales, Robert H. (1990), *Firepower in Limited War,* National Defense University Press, Washington, D.C.

Simpkin, Richard (1985), *Race to the Swift: Thoughts on Twenty-First Century Warfare,* Brassey's Defence Publishers, Oxford, United Kingdom.

Steeb, R., J. Matsumura, T. G. Covington, T. J. Herbert, and S. Eisenhard (1996b), *Rapid Force Projection: Exploring New Technology Concepts for Light Airborne Forces,* DB-168-A/OSD, RAND, Santa Monica, CA.

Steeb, R., J. Matsumura, T. G. Covington, T. J. Herbert, S. Eisenhard, and L. J. Melody (1996a), *Rapid Force Projection Technologies: A Quick-Look Analysis of Advanced Light Indirect Fire Systems*, DB–169-A/OSD, RAND, Santa Monica, CA.

Ullman, Harlan, and James Wade (1996), *Shock and Awe: Achieving Rapid Dominance*, National Defense University Press, Washington, D.C.

United States Army (USA) (1996), *Task Force Griffin: Final Briefing Report*, U.S. Army Training and Doctrine Command (TRADOC), Fort Leavenworth, KS.

United States Army (USA) (1999), *Army Experimentation Campaign Plan*, presentation by COL. Michael K. Mehaffey to the Military Operations Research Symposium, March 8, 1998, at Norfolk, VA, HQ, Training and Doctrine Command, Fort Monroe, VA.

United States Marine Corps (USMC) (1997a), *Exploiting Hunter Warrior*, Marine Corps Warfighting Laboratory, Quantico, VA.

United States Marine Corps (USMC) (1997b), *Hunter Warrior: Advanced Warfighting Experiment Reconstruction and Operations/ Training Analysis Report*, Warfighting Laboratory, 1 August, Quantico VA (limited to U.S. government agencies only).

United States Marine Corps (USMC) (1998a), *Asymmetric Warfare: Future War in the Littorals*, Defense Intelligence Reference Document, Marine Corps Intelligence Activity, Quantico, VA.

United States Marine Corps (USMC) (1998b), *Expeditionary Operations*, MCDP3, Quantico, VA.

United States Marine Corps (USMC) (1998b est.), *United States Marine Corps Warfighting Concepts for the 21st Century*, Marine Corps Combat Development Command, Quantico, VA.

Van Creveld, Martin (1985), *Command in War*, Harvard University Press, Cambridge, MA.